TAIWAN ABORIGINES
A GENETIC STUDY OF TRIBAL VARIATIONS

TAIWAN ABORIGINES

A GENETIC STUDY OF TRIBAL VARIATIONS

CHEN KANG CHAI

THE JACKSON LABORATORY, BAR HARBOR, MAINE

Harvard University Press · Cambridge, Massachusetts · 1967

TO THE MEMORY OF MY MOTHER

"... *So, the way cannot be either heard or seen, but sages, by observing its influences, can infer its semblances.*"
—Lao Tzu, 500 B.C. interpreted by Han Fei Tzu.

PREFACE

The data reported in this monograph were collected during a sabbatical leave from The Jackson Laboratory to the National Taiwan University for the purpose of carrying out a field study of the aboriginal tribes of Taiwan. In the 2 years since the completion of the survey I have been mainly occupied in analyzing the data, which involved extensive calculation. I believe that I have covered the essential points from which certain conclusions can be drawn; therefore the results are presented without further delay.

Often in presentations of original work observations and measurements of individuals are included. I do not think they are important in this study, and I have therefore omitted them as well as the numerical results of calculations that could be stated more clearly and usefully in words. Since in many instances large numbers of variables are involved, I have used diagrams wherever I believe they may be useful in providing an over-all picture, and I have consistently tried to bring them to some focal point.

The monograph is written essentially for geneticists and anthropologists. Some terminology, methods, and procedures familiar to students in one field, but not to those in the other, have been explained and illustrated. Material in the third and fourth chapters, which serves as background information to the study, may be of interest to students of history and demography, since most of it is published in Chinese, some in Japanese, and therefore difficult of access to most of the Western world. The fairly numerous archeologic studies published before World War

II in Japanese, and later in Chinese, have not been cited, since the present work is entirely oriented to biology. However, some biologic papers, which did not seem pertinent as references in the individual chapters, are listed in the bibliography. I believe that they should be included here, since as far as I am aware they have not been comprehensively listed elsewhere.

I wish to express my deep gratitude to the John Simon Guggenheim Foundation for awarding me the fellowship without which this study could not have materialized. The work was also supported in part by Public Health Service Grants CA-03108 from the National Cancer Institute and GM-05550 from the National Institutes of Health and by an allocation from a General Research Support Grant 1 SO1 FR-05545-01 to The Jackson Laboratory, Bar Harbor, Maine.

In preparing this work I first corresponded with Dr. S. L. Chien, President of National Taiwan University. He warmly welcomed me to the University and arranged with Dr. H. Y. Wei, Dean of the Medical College, to make a laboratory available to me. Dr. Wei also helped to secure documentation by various agents to facilitate our field survey. I am grateful to both Drs. S. L. Chien and H. Y. Wei for their help.

My heartfelt thanks and appreciation go to my assistants, Mr. F. S. Han and Mr. T. M. Lin, for sharing the workloads with me, especially in the field trips with their long working days and hard travel in the mountain areas. Their patience and friendly attitude will always be remembered.

I also want to thank the various agents in town and county governments, and the Civil Affairs Division of the Provincial Government, for assistance in arranging our field trips. I thank particularly Miss C. S. Yei, Chief, and Mr. Y. L. Chung, of the Division of Elementary Education, Ministry of Education, Republic of China, who instructed the elementary school personnel in the aboriginal areas to allow us to make tests and examinations, and who kindly gained permission for us to use the intelligence tests, the copyright of which is owned by the Division. I am grateful indeed to the village managers, police officers, elementary school principals and teachers, and health-station medical personnel for their help in arranging space for examinations, in gathering the people together, and for acting as recording clerks and interpreters; I regret that I have no space to mention their names here. And I am grateful to those who came to be tested and examined, each one of whom contributed to the completion of this work.

I am obliged to Dr. E. R. Dempster, Department of Genetics, University of California, for making office space available to me, for consultations regarding analysis of the data, and for making arrangements with the Computing Center of the University for me to carry out the heavy part of the computations; and to other members of the department for the facilities and kindnesses that they provided during my stay at the University.

Finally, I am most grateful to Dr. D. W. Bailey, Cancer Research Institute of San Francisco Medical Center, University of California, and to my colleague at The Jackson Laboratory, Dr. T. H. Roderick, for reading the entire manuscript, and also to those who read individual chapters. Their comments and criticisms have been most valuable and helpful in completing the manuscript. Any errors in this book are my responsibility. I also wish to thank Mrs. Ruth E. Soper and G. C. McKay for making the illustrations, Mrs. Dorothy L. Killam for her editorial assistance, and Mrs. Isabelle M. Stover for typing the manuscript, especially the long tables.

Bar Harbor, Maine

February 10, 1965

C. K. Chai

CONTENTS

I INTRODUCTION   I

2 THE FIELD SURVEY   15

3 HISTORY AND CLASSIFICATION   24

4 POPULATION AND FAMILY STRUCTURE   38

5 ANTHROPOSCOPIC OBSERVATIONS   49

6 ANTHROPOMETRIC MEASUREMENTS   79

7 TASTE THRESHOLD FOR PHENYLTHIOUREA   119

8 BLOOD PRESSURE   130

9 DERMATOGLYPHICS   146

10 INTELLIGENCE   186

11 GENERAL CONCLUSION   209

   BIBLIOGRAPHY   221

   INDEX   233

FIGURES

1. Map showing the villages visited in our field survey   19
2. Map showing the migrations of Chinese to Taiwan   27
3. Topographic map of Taiwan   31
4. Population sizes of the different Taiwan aboriginal tribes from 1906 to 1960   40
5. Population pyramids   42
6. The left eye of a person with the epicanthus without Mongolian fold, epicanthus of a German woman, and a slight epicanthus and complete Mongolian fold of an Atayal woman   56
7. Types of external ears   62
8. A married Atayal couple   65
9. Two Saisiat brothers   66
10. A Bunun woman   67
11. Members of the Tsou tribe   68
12. Members of the Paiwan tribe   69
13. The hands and a foot of the man in Fig. 12   70
14. A Rukai man and a Chieftain of the Puyuma tribe   71
15. Members of the Ami tribe   72
16. Members of the Yami tribe   73
17. A girl of a Chinese father and a Puyuma mother, and a girl of a Japanese father and a Puyuma mother   74
18. The landmarks on the head used for taking the anthropometric measurements   80
19. The landmarks on the body used for taking the anthropometric measurements   82
20. Diagrams plotting the mean head breadth against the mean head length of each tribe   87
21. Diagrams plotting the mean intercristal distance against the mean stature of each tribe   89
22. The relative stature and skin color of a Bunun man with reference to Chinese, and the relative stature of the Paiwan with reference to Chinese   98
23. A scatter diagram showing approximate biologic distances between the tribes   117

24. The results of testing reliability of the PTC taste tests — 120
25. PTC taste threshold with increase in age — 122
26. The distribution of PTC thresholds in each of the eight aboriginal tribes — 123
27. Scatter diagrams showing the distributions of systolic blood pressures at different ages of men of the different tribes — 135
28. Scatter diagrams showing the distribution of diastolic blood pressures at different ages of men — 136
29. Scatter diagrams showing the distribution of systolic blood pressures at different ages of women — 137
30. Scatter diagrams showing the distributions of diastolic blood pressures at different ages of women — 138
31. Diagrams illustrating the distribution of blood pressures of Puyuma women — 143
32. Diagram illustrating the fingerprint types, and the left handprint of a chieftain of the Puyuma tribe — 152
33. Percentage frequency distribution of fingerprint patterns in men of the different tribes — 160
34. Percentage frequency distribution of fingerprint patterns in women of the different tribes — 161
35. Percentage frequency distribution of the palm main line indices for the men of each tribe — 166
36. Percentage frequency distribution of the palm main line indices for the women of each tribe — 167
37. Point diagrams showing the relative "distance" and separation between the tribes — 185
38. Some of the figures used for the discrimination, substitution, and block-counting tests — 189
39. Distributions of the discrimination, substitution, and block-counting subtest scores — 192
40. Two-dimensional space diagram made by plotting the average block-counting test score against the average discrimination test score of each tribe — 204

TABLES

1. Villages covered in the field survey   18
2. The distribution of the aboriginal populations in percentages at various altitudes   34
3. Infant mortality of Taiwan aborigines living in the tribal areas   43
4. Average number of people in a family in the Taiwan aboriginal tribes   45
5. Number of households and people of the different tribes in villages distributed in the tribal areas   46
6. Number of mothers having surviving children in one village of the Bunun tribe.   47
7. Percentage distribution of people with different thickness of the mucous lips   50
8. Percentage distribution of people with different nasal bridge forms   52
9. Percentage distribution of people with different eye apertures   54
10. Percentage distribution of people with or without epicanthus   57
11. Percentage distribution of people with different Mongoloid folds   58
12. Percentage distribution of people with different iris color   60
13. Percentage distribution of people with different types of ear lobe   61
14. Percentage distribution of people with different ear points   63
15. Means and standard deviations of anthropometric measurements for men of the Taiwan aboriginal tribes   85
16. Means and standard deviations of anthropometric measurements for women of the Taiwan aboriginal tribes   86
17. Indices of head and body measurements for men of the Taiwan aboriginal tribes   91
18. Indices of head and body measurements for women of the Taiwan aboriginal tribes   92
19. Correlation coefficients between anthropometric measurements of individuals for all eight tribes   100
20. Correlation coefficients between mean anthropometric measurements of each of eight tribes   101

21. Variance fraction between the eight Taiwan aboriginal tribes for each anthropometric measurement — 105
22. Classification matrix based on the discriminant-function analysis of 19 physical measurements for men — 110
23. Classification matrix based on the discriminant-function analysis of 19 physical measurements for women — 111
24. Classification matrix based on the discriminant-function analysis of 19 physical measurements for men — 111
25. Classification matrix based on the discriminant-function analysis of 19 physical measurements for women — 111
26. Classification matrix based on discriminant analysis of 11 physical measurements for women — 113
27. Biologic distance between tribes based on measurements of men — 115
28. Biologic distance between tribes based on measurements of women — 116
29. Mean PTC taste thresholds of school pupils and adults — 121
30. Distribution of nontasters and gene frequencies — 124
31. Mean PTC taste threshold for tasters — 125
32. Coefficients of regression of blood pressure on age — 134
33. Test of significance of the difference in regression of the diastolic pressure of women — 139
34. Observed and adjusted mean blood pressures of men — 140
35. Observed and adjusted mean blood pressures of women — 140
36. Mean blood pressures at 35 to 39 years of age — 142
37. Frequency distributions of fingerprint patterns and values of χ^2 for men — 154
38. Frequency distributions of fingerprint patterns and values of χ^2 for women — 156
39. Chi-square tests of bimanual and sexual differences in frequencies of occurrence of fingerprint types — 158
40. Whorl-to-loop ratios in fingerprints — 163
41. Average finger ridge pattern intensities — 165
42. Chi-square tests of bimanual and sexual differences in the frequencies of main line indices — 170
43. Average main line indices — 172
44. Total χ^2 for the palmar main line difference between tribes — 174
45. Correlation coefficients of print patterns between fingers — 175
46. Correlation coefficients of the indices between different main lines — 176
47. Classification matrix based on the discriminant-function analysis of 22 dermatoglyphic variables — 179
48. Classification matrix of five aboriginal tribes based on discriminant analysis of the dermatoglyphics of fingers and palms — 180

xvi

49. Mahalanobis' D^2 between tribes based on dermatoglyphics of men   181
50. Mahalanobis' D^2 between tribes based on dermatoglyphics of women   181
51. Mean ages with standard errors of children receiving intelligence tests   191
52. Correlation coefficients between form A and form B of each subtest and the three subtests combined, aborigines and Chinese   195
53. Means and standard errors of the figure discrimination subtest scores   197
54. Means and standard errors of the figure substitution subtest scores   198
55. Means of the block-counting test scores   199
56. Mean scores of each subtest and total mean scores   200
57. Summary of results of analyses of variance for differences between tribes   201
58. Distribution of ABO blood group frequencies   217

TAIWAN ABORIGINES

A GENETIC STUDY OF TRIBAL VARIATIONS

♀♂

CHAPTER I. INTRODUCTION

Problems facing the student of the science of man are multiple and complex. The recent vigorous growth in the knowledge of human genetics, especially in the area of polymorphic traits, has pushed the classical approach to population dynamics and evolution into a corner. Coon wrote in 1962: "In recent decades the pursuit of anthropometry has declined, except for applied anthropology. Instead of measuring the bodies of the last remnants of aboriginal populations, anthropometrists measure military personnel and civilians in order to design railroad and airplane seats and space suits. . . In any case no form of evidence is unwelcome and only by close study of detail can we hope to solve this and related problems." In commenting on past work Washburn (1953) concluded that "the strategy of physical anthropology yields diminishing returns, and finally, application of the traditional methods by experts gave contradictory results." Likewise discouraged, Birdsell wrote in 1950: "The present methodological approaches utilized in race studies are bankrupt." The basic confusion and misunderstanding are due, we are told, to inadequate knowledge of population genetics on the one hand (Dobzhansky, 1962), and to inadequate methods of approach on the other.

In modern science the definitive solution of a problem in any field requires the application of multiple disciplines. Polymorphic traits are unique in the sense that they bear direct correspondence to genes; on the other hand, polygenic

1

characters, which are genetically complex, have been successfully used in systematics and evolutionary studies, and should not be discarded simply because they are not genetically well defined. Each trait has its adaptive significance, each plays a certain role, and each contributes its own bit of evidence; and, if we are to understand complicated evolutionary processes, *all* evidence is important. Physical anthropology is not sterile, but it needs the support of genetic theory in interpreting its observational data and perhaps genetic technics for its approach.

The simplified procedures of classic Mendelian genetics, in which a single gene was studied separately and independently in laboratory populations, were necessary to determine the laws of inheritance and to obtain basic knowledge of the physical and biologic properties of genes. But to deal with many genes acting on the same and different characters and interacting among themselves, and to seek the genetic impact of evolution on characters determined by these genes, it is necessary to study natural populations. Men are more interested in knowing their own biologic history than that of any other creature. And to understand their evolution and genetic differences they must study their own populations.

Since Darwin's theory of natural selection was synthesized in modern genetic terms—the "Darwinism reborn" of about two decades ago (Huxley, 1943)—the development in genetics from genic to populational levels has further advanced. Our thinking in the problems of evolution needs the support of recent knowledge, mainly that gained from study of population genetics.

A species is composed of numerous geographic or local populations, each having a common gene pool shared by all its members, any one of whom has the opportunity of mating and producing offspring. Such a reproducing community of sexual and cross-fertilizing individuals is a Mendelian population—a panmictic unit (Wright, 1931)—a dynamic entity undergoing evolutionary changes.

PHYSIOLOGIC PROPERTIES OF GENES

The physiologic property of a gene has its impact on evolution. In the chemical structure of a gene there is template deoxyribose nucleic acid (DNA) from which are ultimately formed enzyme and structural proteins that determine the development of an organism. Genes do not form the proteins directly, but through the aid of various kinds of messenger ribonucleic acid (RNA) in the cytoplasm. Hence, different levels of gene interaction are postulated (Mayr, 1963). Within a cell several genes act simultaneously to produce a gene product or products that, diffused into other cells, stimulate their growth and division. There is a second level of interaction among the tissues and organs after cell growth is initiated.

What controls the activity of a gene and the site of its activity are still largely unknown. We do know that a gene which aids fitness in one genetic background may be deleterious or even lethal in another, and that an organism to survive and produce offspring must have a harmonious genotype. Thus, in addition to the cellular and developmental levels there is a further level of interaction among genes, in which the fitness of a gene depends upon the kind of "gene team" with which it is associated in the production of a particular phenotype. It is evident that the goodness of a gene is statistical, averaged from all the different genetic backgrounds in the many genotypes in which the gene is exposed to selection, and that the selective value of a gene is the mean of the selective values of all the combinations in which it appears in a given population (gene pool).

Pleiotropy, the property of a gene affecting the different

characters of an organism, is significant in evolution. Since the primary gene action in multicellular organisms is several steps apart from the expression of the peripheral character, genes without pleiotropic activity must be rare. The importance of pleiotropy in evolution was recognized by Chetverikov as early as 1926 and emphasized later by Dobzhansky and Mayr, and others.

The biologic properties of the genetic determinants of quantitative variation are still confusing. The term polygenes was proposed by Mather (1943), who postulated that polygenes are the determinants of quantitative characters. Their individual physiologic effects on a given character are small but additive and supplementary. Concerning the genetic units of quantitative characters Wright (1952) commented that "the evidence that the genetic component of quantitative variability depends on multiple factors distributed through the chromosomes seems at present adequate. . . Satisfaction with this theory has, however, fallen short of completeness." Mayr (1963) recently asked: "Are the small phenotypic effects of quantitative inheritance caused by genes whose only function is to be specific modifiers, or are they the pleiotropic byproduct of genes having other essential functions in the organism? It is of great importance in evolution what alternative is right." Concerning Mather's polygene concept Mayr has commented that it tends to ascribe to each gene a specific and singular function and is similar to the modifier concept of the early geneticists to which Goldschmidt subscribed. These concepts are essentially based on one gene—one character hypothesis. Mayr claimed that a gene major in its own right may serve as a modifier of other genes, hence as a polygene for other characters. He used the term polygeny to describe this property. Be that as it may, the theory of polygeny contributes to our understanding of evolution.

EVOLUTIONARY MECHANISMS

Factors that contribute to variations within and between populations are multiple. Besides natural selection there are also migration, mutation, drift, and social and geographic isolation, which cause and condition genetic changes in natural populations.

Natural selection, mutation, and drift. Natural selection tends to bring about the most favorable composition of a population's gene pool, in which the genes are adaptive not only to the environment but also to each other (Dobzhansky, 1962). The emergence of such a coadapted gene system in a population is thought to be mainly a consequence of two kinds of natural selection: (1) normalizing (Waddington, 1957) or stablizing (Schmalhausen, 1949), which tends to limit the extent of deviation from the population mean, and (2) directional selection, which tends to move the mean. The two forms of selection coexist in all natural populations.

In natural populations mutation provides the raw materials of evolution, and natural selection uses them to build organisms adapted to their physical and social environment. The effect of mutation on quantitative traits is not easy to demonstrate, especially in natural populations, for phenotypic changes caused by mutation cannot be distinguished from those caused by recombination of the genes affecting the same trait. Genetic drift is possibly another factor causing variations between isolated groups. Ford (1964) pointed out that in natural populations selection is much more important than drift. This would be so particularly for characters determined by a large number of genes and highly adaptive.

Closely associated with the theory of natural selection is the recent controversial issue regarding "genetic load," for which there are two hypotheses: classical and balance

(Dobzhansky, 1955). The classical hypothesis implies that evolutionary changes consist in gradual substitution and eventual fixation of the more favorable alleles, superior gene alleles being established. Inferior ones, deleterious in heterozygotes as well as in homozygotes, the so-called mutational load, are eliminated by natural selection. The balance theory implies that, in addition to the mutational load, the adaptive norm is an array of genotypes heterozygous for many gene alleles and that the homozygotes for these genes are in a minority and inferior in fitness, so that natural selection leads only in part to gene substitution. Distinguishing between these two hypotheses and expressing his own opinion, Dobzhansky (1962) said: "At a risk of oversimplification, it may be said that many of the genes that cause major hereditary diseases and malformations probably belong to the first class. The genes that produce differences of the sort we observe among healthy and 'normal' people might fall mostly in the second category. The genes conditioning the variations in the appearance, physique, intelligence, temperament, special abilities—in short, the genes making people recognizably different persons, really unique and nonrecurrent individuals—may be maintained in human populations in balanced states, either because of being advantageous in heterozygotes or because of the action of diversifying selection."

Sexual and social selection. In our complicated human society some economic and social factors in selection are detectable and others are not, varying between societies or populations. A gentleman may prefer a blonde; a huntsman, a mate with curly black hair. The choosing of mates may be affected to a great extent by the family economic level. Sexual selection may be different between one ethnic group and another, and in time and space. In the highly developed consciousness of human beings, whether primitive or civilized, selection for mates definitely plays an important role in evolution and genetic changes, especially in the morphologic and behavioral traits. Furthermore, marriage customs, family structures, and population size all contribute to genetic variations. Human populations are most imperfect as Mendelian populations, and they deviate greatly from ideal panmictic units.

GENETIC BASES OF EVOLUTION

To understand fully the genetic mechanisms involved in evolutionary changes, the nature of inter- and intraspecific differences needs to be examined first, and then the way in which variability and response to selection in populations have been maintained. Genetic characters may be classified into two categories: one, affected mainly by single or small numbers of genes, giving a discrete variation; and the other determined by polygenes, giving a continuous variation. Interspecific differences are due more to polygenes than to major gene differences, since similar variants are often found in related species, genera or even higher groups (Haldane, 1925; Mather, 1943). Similar types of heritable differentiating variation are also found within species. But within species there are two kinds of major gene variations, one maintained in balance, namely, polymorphism, the other unstable, including the lethals, sublethals, and various structural aberrants. Some of the latter types, reducing the viability of the carrier, have no evidence of specific difference and hence no clear evolutionary significance. Polymorphism requires a balance of selective advantages for maintenance of the variation in the population.

Genetic determinants of importance for studying evolution and racial relationships are the polymorphic genes and the polygenes. In classifying living races, one school uses blood groups and the other fossils and head and body forms.

Poor understanding of the behavior of these genes is a basic source of controversies and discrepancies regarding studies of racial relationships.

Polymorphic variation. Ford (1940) defines polymorphism as the simultaneous occurrence in one habitat of two or more distinct forms of a species "in such a proportion that the rarest of them cannot be maintained by recurrent mutation." The maintenance of polymorphism by selection in favor of heterozygotes was hypothesized first by R. A. Fisher and later by Ford, who drew such supporting evidence from his own studies in ecologic populations and from the work of others. Ford (1962) used the supergene theory (Darlington and Mather, 1949) to explain a switch mechanism in controlling the balance of polymorphic genes. According to him the formation of supergenes is by chromosomal inversion or translocation resulting from natural selection in favor of the interacting genes concerned, so that they are transmitted as a single unit. He has pointed out many other genetic mechanisms for maintaining polymorphisms, but he seems to believe that the majority of them are controlled in stable equilibrium by supergenes or other switching mechanisms. When polymorphism is maintained by selective advantage of the heterozygote, this genotype should exceed its expected frequency of the Hardy–Winberg law in a stabilized population (Williamson, 1960). But in many cases the genetic basis for the coexistence of two or more different gene alleles in a population is not clear. Li (1955) demonstrated that when two or more different alleles are favored by selection in different spatial and temporal environments, the coexistence of genotypes within a Mendelian population is comparable to the coexistence of isolated reproductive species. Another kind of balance of polymorphism has been considered by Wright and later by Lewontin (Dobzhansky, 1955), e.g., the selective value of a genotype decreases as its frequency increases. That is, an allele is advantageous when it is rare but disadvantageous when it becomes common. Perhaps in general both the gene pool of a population and its environment are important in maintaining a given balance for a polymorphic trait, although this may not be easy to prove. In the human species, there is a significant excess of heterozygotes of MN progeny from the mating of MN × MN. The well-known incompatibility in the Rh blood groups results in hemolytic disease. The Rh-negative gene is selectively disadvantageous when its frequency is low. There is no obvious indication that the heterozygotes are favored by natural selection. It has been reported that a significantly high proportion of individuals developing stomach cancer belong to blood group A (Aird et al., 1953). It has also been shown that the carcinogenic stimulus is not directly related to A-antigen secretion by the salivary glands (Clarke et al., 1955). The etiology of these cancers is genetically complex. It has generally been assumed that many genes determine the susceptibility. These studies suggest the association of a blood group with genes causing the development of disease. The traits on which natural selection acts are not the blood group antigens so far as evidence to date has shown. There are examples of association of polymorphic traits with adaptivity in some other species as well. For instance, variability of color in the corn weevil *Bruchus quadrimaculatus* does not have any protective value. However, color variability is associated with great differences in fertility: weevils of normal coloration are twice as fertile as the black type and almost twice as fertile as the red, whereas gray weevils are one-twelfth as fertile as the normal ones (Schmalhausen, 1949, p. 165).

Thus it appears that different traits have different controlling mechanisms. Some are associated with simple, but most with rather complex, genetic traits. It is these associated traits, or genetic complexes, that are of adaptive significance. The following genetic mechanisms may account for the association: (1) pleiotropism, (2) linkage, or (3) integrated gene pool—as a member of a coadapted gene pool, the gene frequency is affected by the totality of the composition of the genes in a population. It may generally be said that the controlling mechanisms of polymorphisms vary among traits. Those that show an excess of heterozygotes may have relatively simple mechanisms, whereas those that do not may have rather diffuse control mechanisms and be genetically related to many traits or to the whole integrated gene pool of the population.

Thus two different races or local populations with a widely separated genetic relationship may have rather similar frequencies for certain polymorphic genes, whereas two closely related geographically may have different balances for special polymorphic genes. In the blood groups of the human species the Rh gene frequencies show racial or geographic (continental) differences (Mourant, 1954). In the ABO and MN blood groups, gene frequencies are also different between major races or groups, but the differences are slight and there is much overlapping. Furthermore, there are exceptions that are difficult to explain on a racial basis. For instance, it is generally agreed that the American Indians are derived from the Mongoloid race, and yet North American Indians have O and A, but no B and AB blood groups, whereas the Chinese and Japanese, who are subraces of the Mongoloid, have rather high percentages of B and AB.

Polygenic variations. Mather (1943) mentioned that a few decades ago Galton, Pearson, and their associates disclosed a wealth of biologic variation in man suggestive of polygenic interpretation, and nearly all important characters, such as yield, quality, and disease resistance in domestic animals

and plants, are polygenic. Thus naturally occurring polygenic variations are widely spread within and between
species. In man these include all anthropometric measurements, skin color, hair forms, some physiologic traits, and
even intelligence. Concerning their significance in evolution Mather said: "Polygenic inheritance is marked by certain peculiar features which distinguish it from oligogenic
behavior and which throw a fresh light on the interrelations
of variation and selection. Polygenetics represents a new
level of integration by means of which a better understanding of natural selection and its action may be achieved."
Mayr's concept of the action of polygenes as previously
mentioned naturally does not invalidate Mather's hypothesis
but complements it.

The characters that have long been used by anthropologists in studying evolution are mainly polygenic. Indeed, in
natural populations practically all normal variations except
polymorphic traits are of this type. Owing to the complex
genetic mechanisms of these traits, their value in studies of
evolution and racial relationship has not been fully
appreciated.

The adaptability of many polygenic characters has been
demonstrated in various species either by experiment or observation. Body size in animals or stature in man, representing a total of mean measurements of many skeletal parts of
the body, is a complex trait. The genetic basis is complicated but it is partly hereditary and determined by many
genetic factors or polygenes (Wright, 1932; Chai, 1956a). For
instance, an increase of body size involves a more econom-
ical metabolism, and this is true for all organisms (Schmalhausen, 1949). Thus large body size as a polygenic trait in
birds and mammals should have great physiologic significance to adaptation in cold regions. Bergmann's rule is that
body size of animals in related groups increases with tran-

sition from a warm to a colder climate. Under laboratory conditions it has been found that extremely small body size is not genetically favored: a strain of mice known as Small, characterized by very small body size, drifts continually upward in body size, whereas the Large strain, characterized by very large body size, remains rather constant (Chai, 1961).

Another example is dental characteristics. Teeth are considered most important in fossil studies in paleontology and speciation of mammals. The remarkable evolutionary changes in the teeth in the human and other primate species have been well illustrated by Coon (1962). The ordinary variation in number and shape of teeth, as ordinarily observed in populations, is determined by polygenes (Lasker, 1950). The lack or incomplete development of the third molar in mice (Grüneberg, 1952) is comparable genetically to that in man (Grahnem, 1961). Garn et al. (1963) believed that the third molar agenesis is associated to agenesis of other teeth and to delayed formation time of the remaining posterior teeth. This type of phenotypic variation, a so-called threshold character with or without various degrees of expression, is quite often observed during the development in the organisms of different species. The polydactyly in guinea pigs (Wright, 1934) is a classical case, as are rumplessness (Landauer, 1955) and diplopodia (Talor et al., 1959) in fowl; and "zigzag" (Lyon, 1960) and "careening" (Chai and Chiang, 1962) conditions of locomotion in mice. Although there have been one or two factor hypotheses for explanation of certain variations of this type, the genetic basis is in general more likely to be more complex, involving possibly many genes with individual effects of *different* magnitudes. It is possible that in this type of variation the number of genes involved may be fewer and their individual effects stronger in comparison with characters that can be linearly measured, such as size, yield, and performance.

On the other hand, the fact that polygenic traits have the properties of *homeostasis* and *flexibility* is due not only to the large number of genes responsible for the traits as Birdsell (1950) pointed out, but also to the very properties of the polygenes, such as additive effects, which bring about gradual populational changes in response to natural selection, and a relational balance between linked genes, which is important for maintaining heterozygosity and some intermediacy in outbreeding organisms (Mather, 1943; Bodmer and Parsons, 1962). Thus Mather (1943) said "The successful survival of a species depends on maintenance of a balance between free and potential variability. . . . A complete absence of heritable variability would render the population incapable of future adaptation, except by mutation, and so doom it to extinction." Mather's theory on the flexible properties of the polygenic traits is essentially concerned with variability potential. It is not feasible here to state his theory fully.

Other properties concerning polygenic characters are pleiotropism and linkage. For a large number of genes involved in the quantitative characters these properties are important in their response to selection. Correlations exist between many polygenic characters as shown in our correlation analysis of the anthropometric measurements and of the fingerprints and the main ridge lines in the palms. When two or more characters are correlated through such genetic mechanisms, usually one is favored by selection, and under certain conditions most of the other correlated characters may nullify the selective advantage of that character, as environmental conditions are not likely to favor all the characters concerned (Mather, 1943). The result of this is a slowing down of genetic changes, but gradual changes proceed step-by-step through a reshuffling of the genes toward the optimum, and a slow divergence between the populations follows. This is another essential mechanism by which species and subspecies retain the original forms and shapes characteristic of their early ancestors through many generations.

The following data were collected in a field survey among fairly primitive groups—the Taiwan aboriginal tribes, which are social as well as geographic isolates. The data cover tests, measurements, and observations of both polymorphic and polygenic traits in individuals of the different tribes. The genetic variations built up in these tribes were the primary inquiry of this study.

In treating the data we used the χ^2, t, and F statistics, regression, correlation, and variance analyses, discrimination function, and Mahalanobis' general distance analyses for handling large numbers of variables. In interpreting the results we took into consideration some historical and archeologic evidence, social practices, family structures, population sizes, and geographic variations, as well as current theories in population genetics and evolution. After a brief look at the early history of the Taiwan aborigines, and a review of the population size and distribution of the present tribes, an attempt will be made to establish their biologic relationship from the over-all evidence. The data collected in our survey provide the basis of my position regarding the evolutionary mechanisms involved and the relative significance of the various traits in studying racial relationships and evolution.

♀♂

CHAPTER 2. THE FIELD SURVEY

Early in September 1962, when I arrived in Taipei, I set about solving my first and most pressing problem, the organization of a work team for the field survey. I hoped first of all to find a medical man who would be willing to join me, and I spent three hot, humid weeks—it was the typhoon season—making inquiries to that end. The outcome, however, was completely discouraging; there seemed to be no qualified physician who wanted to undertake the task. This, though disappointing, was quite understandable, since a physician would not only risk damaging his practice by a protracted absence but would receive only a negligible income during it. I therefore gave up the idea and started searching instead for a biologist.

I was most fortunate in finding Mr. F. S. Han willing to join me. A graduate of the National Taiwan University, with degrees in animal breeding and genetics, Han was at that time preparing for advanced training, which he later undertook in the United States. He was greatly interested in genetics and in the proposed survey, and proved to be a hardy and persistent worker.

We two were then joined by Mr. T. M. Lin, a teacher's college graduate working in the Department of Education in Taipei. Experienced in giving intelligence tests to elementary school children on the island, accustomed to working with the people, and well grounded in all that is involved in Chinese documentation, Lin was an invaluable addition to our team. He also, I was delighted to find, was willing to take over the duties of technical assistant and secretary, attending to such matters as our rather extensive correspondence with various government agencies.

I was confident that with the cooperation of Han and Lin the study would be well implemented. Later on we were joined by Dr. David Wang, a dentist; I regret that Dr. Wang had not completed his dental survey at the time I left Taiwan.

During those first weeks in Taipei we gathered and prepared all the necessary equipment, made PTC testing solutions, and had various forms printed for recording anthropometric measurements, intelligence test data, and dermatoglyphic prints. We wanted to perfect the technics of making anthropometric measurements to be able to do them quickly and accurately in the field, and Han and I were kept busy practicing on various people and on each other, aided by the staff of the Anatomy Department of the Medical College. We are indebted to Professors C. C. Yu and T. M. Tseng for their good advice and for the loan of anthropometric equipment.

Our field trip began October 23, 1962. Starting from Taipei in the northern part of the island, we traveled along the west side to the southern tip, where we turned and went back up along the eastern side. This schedule took advantage of the fact that though northern Taiwan is chilly and rainy in winter, it is warm and pleasant in the south. Following this route we visited the tribes in the following sequence: Atayal, Saisiat, Bunun, Tsou, Paiwan, Rukai, Puyuma, and Ami. In 24 villages and 20 public schools we examined and tested about 1900 adults and 2600 school children.

Along the way we received the assistance of medical personnel at the various health stations, and of the village managers, school principals, teachers, and policemen, who served as interpreters and clerks, and helped get the aborigines together for examinations and tests. Since all these people were provincial or government employees, we had to request permission of various agencies at provincial and local levels to obtain their assistance. Also, although we had been informed that head-hunting had been completely abandoned, and that the aborigines were now very peaceful citizens, we found that police permission was still required to go into and out of some tribal areas, and we accordingly applied for the necessary permits, and later found them not only reassuring but immensely helpful in our work. I was pleasantly surprised by the efficiency with which documents were executed, and by the willingness of the local public officials to help us. Their attitude was most friendly, and for this I am grateful.

We sent official documents ahead of us to the county government officials for each aboriginal tribe, to notify them of the date of our arrival, and often an officer of the county government met and escorted us to the village manager, health-station medical personnel, policemen, and school officials. On the day we arrived, or the day after, a planning conference was held with those in charge of administering aboriginal affairs. At these meetings our time schedules for visiting the various villages, and the routes by which we would travel to them, were laid out. Ordinarily we selected the larger villages and some smaller adjacent ones, since the people in these villages were recommended as typical of their tribes.

The villages we visited in each tribe, together with the counties and prefectures to which they belong, are listed in Table 1 and shown on the map (Fig. 1). Some very small villages, where only a few people came to be examined, are omitted.

Table 1. Villages covered in the field survey.

Tribe	Villages	County	Prefecture
Atayal	Wulai, Chungchih,	Wulai	Taipei
	Chunyang, Wushei	Jennai	Nantou
Saisiat	Tungho, Nanchiang, Fenglai	Nanchiang	Miaoli
	(Hungmaokuan)		
Bunun	Lona, Tili	Sein-I	Nantou
Tsou	Tapan, Loyeh	Wufeng	Chia-I
Paiwan	Santi	Santi	Pingtung
	Chiaping, Chungtsen	Taiwu	Pingtung
	Kuloh	Lai-I	Pingtung
Rukai	Chungtsen	Taiwu	Pingtung
	Awu	Santi	Pingtung
	Tewen	Painan	Taitung
Puyuma	Malan		Taitung
	Nanwang, Chihpen	Painan	Taitung
Ami	Malan	Painan	Taitung
	Lungchang, Hsinchang	Tungho	Taitung

Intelligence tests were conveniently given to children in the local schools; examinations and tests for adults took place either in school classrooms or in the health stations. Although at first we tried making tests and measurements in individual homes, we found that this was not practical because of the amount of time and space needed to set up the equipment, and also because of the extremely primitive conditions in some areas; in addition to the complete absence of any kind of sanitation, the one-room dwellings were often crowded not only with many adults and children but also with poultry and livestock, making it exceedingly difficult to carry out our work.

The adults were gathered together and instructed by the village managers, with the cooperation of the school teachers, to come to the appointed examining place. In this way we were able to handle groups of 20 to 40 in half a day.

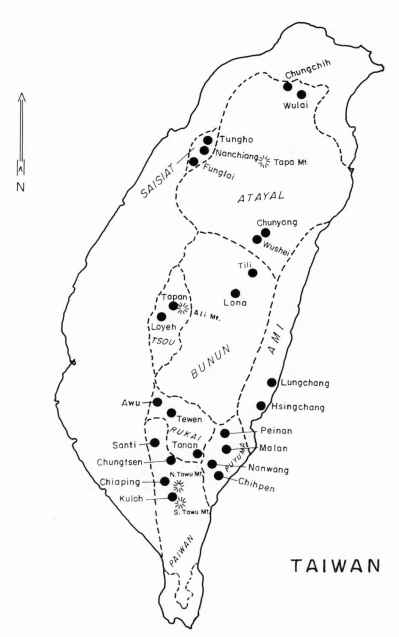

Fig. 1. Map showing the villages visited in our field survey.

Individual desks were set in separate rooms along with the necessary equipment. Each individual's date and place of birth, and ancestry back to his grandparents, were first recorded. Examinations were then given in the following sequence: anthroposcopic observations, body weight, PTC taste tests, blood pressure measurement, dental examination, anthropometric measurements, and finally dermatoglyphic prints of hands. The technics for each examination and test are described in the individual chapters devoted to them. Anthroposcopic observations, PTC tests, dermatoglyphics, and some physical measurements were made for school children as well as for adults in some of the tribes at the beginning of our survey, but later, because of lack of time, these had to be discontinued.

The intelligence tests were given by Mr. Lin to all the school children we examined in each tribe except the Yami, in which the tests were administered by the local school teachers. As the ship from Taiwan to Lan Hsu island, where the Yami live, sails only biweekly and, when the water is rough, does not sail at all, we were unable to make this trip. I regret that our study of the Yami people was limited to the giving of intelligence tests, for which the local school principal came to Taipei for instruction by Lin in methods and procedures.

One problem that plagued us throughout the trip was the lack of adequate transportation. Neither the Medical College nor the National Taiwan University had a vehicle to lend us for the survey, and we found it impossible to get around entirely by public conveyance. When there were buses to rural areas, they were sparsely scheduled and consequently very crowded, as we discovered on our first trip to Wulai, laden with luggage; we could barely squeeze ourselves and our equipment into the bus already tightly packed with people carrying all sorts of bundles, live chickens, etc. Some areas were too remote for bus service but might have truck routes reaching close to, if not into, the villages. But sometimes there was not even a truck route, so that a 3- or 4-hour walk was required, as in our trip to the village of Tapan in the Tsou tribe.

To reach this village we traveled first by train from Taipei to Gu-I, where we boarded another train for Shi Tse Lo. The little narrow-gauge railway train, consisting of a locomotive and two cars, one for passengers and one for lumber etc., took 5 hours to cover a distance of 40 miles. This slowness was partly accounted for by the switchback method by which it climbed the mountain: the locomotive would pull the cars to a sharp-angle turn in the railbed, where it would then go out onto a spur track, reversing so that on the next lap it would be pushing instead of pulling the cars. By repeating this alternate pushing and pulling, the steep zigzag up the mountainside was traversed. From Shi Tse Lo there was no way to travel the remaining 10 miles to Tapan except by footpath through the steep and rocky terrain. So we set out, Han, Lin, two persons hired to carry our luggage, the local school principal who had come to meet us, and I.

A special feature of this walk was the crossing of three of the rope bridges which, spanning the distance between opposite mountainsides, hang fragile as spiderwebs above deep gorges whose rough floors lie 300 to 500 feet below. That these bridges save the tribal people tremendous effort and many hours of time there is no doubt, since without them they would have to go all the way down one mountain and up the next; but to an outlander unused to them they can be quite intimidating. Such a bridge consists of two long parallel rope cables which, bolted at either end to rocks, hang suspended like a hammock, supporting between them, by means of a network of wires, a walkway of boards an inch thick, about 4 feet wide, and extending up to a block in length. When walked upon, the whole structure sways from side to side and jounces up and down. Unfortunately, since the cables are spread at an angle *outward* from the walkway, the person crossing, swaying and bouncing at this dizzy height, has absolutely nothing to hang on to!

We stayed 2 days in this isolated village, sleeping in the home of the school principal, all in one room, on floor mats; and eating salt pickles and rice. (Ordinarily we carried canned food and bread with us, but in this case did not because we had been told that we could obtain meals in the village.)

In spite of our transportation difficulties we managed to get around by using a combination of personally borrowed automobiles, public conveyances, taxis and, in some cases like the above, our feet. We spent about 4 months traveling in the tribal areas and back and forth from them to our home base in Taipei. While among the tribes we usually got up at daybreak and worked until sunset. I was fully satisfied with the quantity and quality of the work achieved during this short period.

When I arrived back in the United States on March 1, 1963, my first task was to organize the data in a form that could be keypunched onto IBM cards for statistical analysis, and to read and analyze the dermatoglyphic finger and palm prints. We had kept well abreast of the intelligence tests while in the field; read by persons who proved to be careful and competent, these were completed soon after we finished the field trip. Two months later we completed a preliminary reading and analysis of the dermatoglyphics, according to the methods of Cummins and Midlo (1961).

After a month spent scrutinizing and checking the records, and discarding those that for some reason were incomplete or of doubtful accuracy, I designed appropriate codes for transferring the data to IBM cards, and they were copied onto roster sheets for keypunching. This gave me another chance to check errors; unfortunately it also gave me a chance to create them, and for this reason I asked my wife, Ling C. Chai, to help me copy the data. This I believe has been most carefully done, and I am grateful to my wife for sharing this tedious task with me. Although some rechecking was occasionally necessary during analysis, the data were now in a form ready for keypunching onto the IBM cards. The heavy statistical analyses were made by using the computing facilities of the Computing Center, University of California. The remaining analyses, and the writing for publication, were carried out after my return to The Jackson Laboratory on September 1, 1963.

♀♂

CHAPTER 3. HISTORY, CLASSIFICATION,
AND DISTRIBUTION

Taiwan is an island of peapod shape, 90 miles across at the widest part and 230 miles long, situated in the China Sea at 22° to slightly beyond 25°N latitude and 120° to 122°E longitude. (It is known also as Formosa, the name given to it by the Portuguese around 1500; however, since the Chinese name of Taiwan, established in the latter part of the Ming Dynasty (Kao, 1958), is still the official title, it is used throughout this book.) Taiwan's closest neighbor is the mainland of China; the coast of Fukien Province is only 90 miles to the west. The Philippine Islands, 180 miles to the south, are next closest. Farther away are Korea and Japan to the north; Indochina, Malaya, and Indonesia to the southwest; and to the southeast New Guinea and many small islands inhabited by Melanesians and Micronesians. This separation by water forms a natural barrier between the people of Taiwan and the surrounding races and ethnic groups.

Within the island the Snow Mountain and Central Mountain ranges, with peaks up to 9000 feet, separate the east and west sides. The rivers and valleys on both sides of the mountains drain into the Pacific Ocean, forming the geographic niches that kept the early tribes isolated.

The average rainfall in Taiwan varies from 4 inches in some coastal areas to around 200 inches in the mountain valleys. The climate is subtropical; only on the high mountain peaks do snow and ice ever accumulate in winter. Vegetation covers practically the whole island, from the cultivated low land to the mountains, which are densely clad with trees, shrubs, and wild grass. In addition to both fresh and salt water fish there are various game animals in the mountain regions. These rich natural food resources were undoubtedly important to the survival of early indigenous tribes and also to successful occupation by invaders.

The enthnography of any country is shaped first by its geography and then by its history of invasion and settlement. The more remote and less desirable areas are usually the last to be populated, serving as places of retreat for early primitive stock when threatened by invaders. So the geography of Taiwan may be the most significant factor in the continued separate existence of its indigenous peoples: the mountains into which they were forced, the valleys and rivers, all served as physical barriers between individual tribes and between the aborigines and the Chinese.

According to Kao (1958) the earliest written record of Chinese invasion of Taiwan is 230 A.D., during the period of The Three Kingdoms, when Sun Chuan sent 10,000 men who fought the Ping Pu tribe (described later) and won. During the Suei Dynasty (610) Chinese soldiers were again sent to Taiwan. There is a possibility that after each war some of the soldiers settled on the island.

During the Soong Dynasty (1225) the Mongolian invasion of China forced many Chinese to flee from the mainland to Taiwan (Chen, 1954). This was possibly the largest emigration of Chinese to Taiwan up to that time. In 1371, in the period of the Ming Dynasty, emigration to Taiwan was forbidden.

The Portuguese visited Taiwan in 1498, but no record can be found of how many landed, or where.

In 1610 a man from the Fukien Province, Go Si Chie, went to Japan and became a merchant, sailing his boats along the east coast of China to Japan. He was joined by Cheng Chi Lueng, and the two together became so powerful that they controlled northern Taiwan between 1619 and 1624. In 1626, during a period of starvation in the mainland of China, Cheng transported thousands of refugees from Fukien to Taiwan (Chen, 1954).

Both the Dutch and the Spanish occupied Taiwan between 1624 and 1642. Then the island was recovered by China following the defeat of the Dutch by General Cheng Chung Gung, son of Cheng Chi Lueng. After the Sino-Japanese war Japan occupied Taiwan, from 1895 to 1945, when it was recovered by China a second time. Within a few years following the end of World War II thousands of Chinese civilians and servicemen moved to Taiwan, the largest emigration of Chinese to the island in history.

Figure 2 (Chang, 1959) shows where the various Chinese emigrations to Taiwan originated, the areas they first settled, and later expansions resulting both from population growth and from the addition of new immigrants during the Ming (1368–1643) and Ching (1644–1911) Dynasties. Note that the Chinese settlement of the east side of the island took place only during the Ching Dynasty, much later than the settlement of the west side. The dates at which Chinese immigrants settled various regions are important, since after those dates it is unlikely that there would have been immigrations of people other than Chinese into the same regions.

Although there have been recent occupations by Dutch, Spanish, and Japanese there are practically no immigrants or descendants of immigrants of these nationalities living as civilians on the island.

The first mention of the existence of indigenous peoples

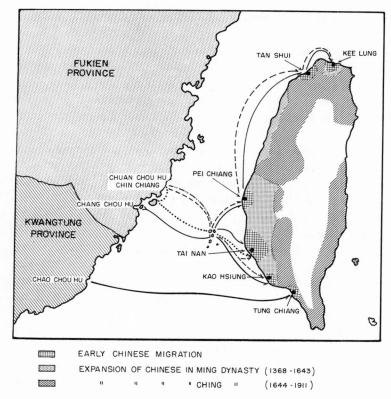

Fig. 2. Map showing the migrations of Chinese from different localities along the coast in the southern part of China to Taiwan in different periods, and the areas where they settled.

in Taiwan is in The Shoo King, the earliest Chinese history. One chapter, written by Yu Gwung, describing events claimed to have happened 4000 years ago, mentions that the people on an island off the mainland of China wore linen clothes and brought bamboo utensils and decorative sea-shells to the Chinese emperor as gifts. This island is claimed by Kao (1958) to have been Taiwan. The earliest actual description of the aborigines and their cultures is found in a Chinese geography written by Jung Sheng in the period of The Three Kingdoms (around the 3rd century A.D.). Following is an abridged translation of a section of this book:

"Yi Chow [Taiwan] is in the sea about two thousand *li* [a unit for measuring distance adopted in China] from the mainland, where there is neither frost nor snow; plants grow the year round. There live the mountain people of different groups; each group occupies a given territory and has its own leader. The people have the custom of piercing their ears and wearing their hair short. Yi Chow is a rich territory, having grain in the land, fish in the water, and minerals, such as copper and iron, in the mountains . . .

"All the people in a family sleep in one bed . . . They make fine patterned cloth. They use deer horns to make weapons, and blue stones for arrowheads and knives, and decorations for women. The people preserve raw fish in barrels for a long time and consider it a delicacy. For getting people together for any event they beat a hollow box which sounds like a drum. In festivals and celebrations people are assembled in a public ground. Five to ten people eat in a group; fish is served in wooden containers and wine in bamboo tubes. They sing and dance . . .

"When they get human heads they remove the brains and the soft tissues and decorate the heads with dog hair, and put clam shells in the mouths. The chieftains of the groups wear them as masks in battle. They bring back human heads which they get in war and hang them on tall poles in their halls; the more they collect, the greater their merits in war . . .

"The parents arrange the marriages of their children. The sons and their brides stay with the parents. Married women have one of the upper front teeth removed."

The chief significance of this source to the present study is that it reports the existence of separate social or breeding groups before the 3rd century. The lineage of the present groups from these is difficult to trace, although in view of the differences in language and culture among the present aboriginal tribes it is very possible that they derived from the ancient ones. But convergence and divergence among the groups, and the extinction of some, have undoubtedly taken place over the long span of time.

Most students of anthropology in Taiwan agree that the aborigines belong to Southern Mongoloid or Oceanic Mongoloid stock; this inference is based on the fact that in some respects their languages, cultures, and body forms are characteristically Indonesian or Malayan. But since there are no reliable records of the earliest migrations of various populations to Taiwan, ethnic relationships are a matter of speculation.

In his study of the early history of the Taiwan aborigines Ling (1954) mentioned that Torii suggested a relationship between the Taiwan aborigines and the Miao people on the mainland of China, and claimed that the Taiwan aborigines are a mixture of Miao and Malayans. Ling himself, after many years studying the indigenous peoples in the southwestern part of China, came to Taiwan and found similarities in culture between the aborigines and those mainland tribes. He refers to the fact that in ancient times there were Yuei Pu tribes (Indonesian) living south of the Yangtze river, some on the east coast, called Pai Yuei, and some in the mountain areas of the southwest, called Pai Pu. Ling claims that the ancestors of the Taiwan aborigines were the Pai Yuei who migrated from the east coast of China. The Yuei Pu people left on the mainland were influenced by other ethnic groups, but they retained various amounts of their original culture, some of which is in common with

that of the Taiwan aborigines. (According to Ling, Naka-
yama (1946) reached a similar conclusion from his study of
the customs of tooth extraction.) Ling further mentions the
possibility of a later migration of people from the Philippine
Islands to Taiwan; he suggests that these people may have
either mixed with the early settlers or established their own
tribes.

Kano (1955) also found a close cultural relationship between the Taiwan aborigines and some ancient tribes on the
east coast of China. At the same time, he maintained that
the culture of the aborigines on the east coast of Taiwan is
saturated with the "Island culture" of the South Pacific,
and on the west coast with the "Continent culture" of
South Asia. According to Kano, some investigators believe
the flow of culture to have been southward from the Continent to the South Pacific islands, through Indochina or
the Malay Peninsula; others believe it was through Taiwan
to the Philippine and other islands. Kano further stated
that in the deep mountain regions of Indochina he found
aborigines who speak the "Island languages." Since the
language of some of the Taiwan aborigines has similarities
with "Island languages," he thought there is a possibility
that some of them may have come from Indochina.

During the Ch'in (255–231 B.C.) and Han (231–220 B.C.)
Dynasties of China communication between Taiwan and
the mainland was forbidden. This fact seems to imply that
if the ancestors of the Taiwan aboriginal tribes did come
from the east coast of China, as Ling claims, they very
likely came before 200 B.C.

There are now nine groups of aborigines inhabiting
Taiwan, eight in the mountain regions and one on the little
island of Lan Hsu (Fig. 3). They are Atayal, Saisiat, Bunun,
Tsou, Paiwan, Rukai, Puyuma, Ami, Yami. This is the
official classification made by the Chinese government in

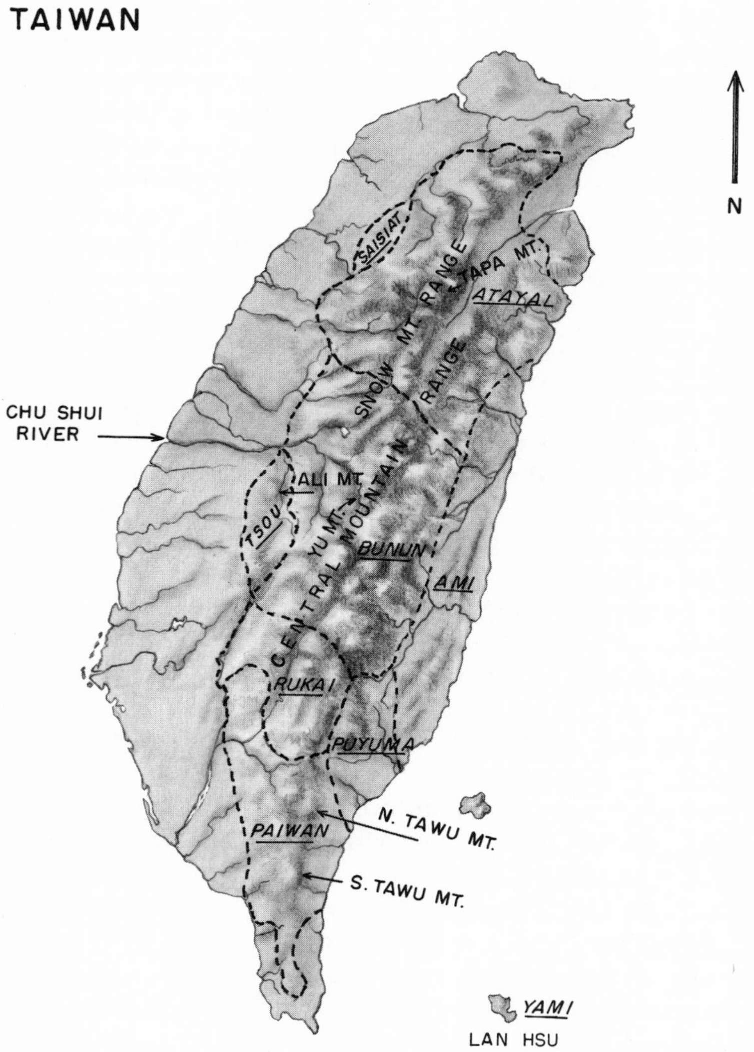

TAIWAN

CHU SHUI
RIVER →

SAISIAT

SNOW MT. RANGE

TAPA MT.

ATAYAL

ALI MT.

TSOU

YU MT.

CENTRAL MOUNTAIN RANGE

BUNUN

AMI

RUKAI

PUYUMA

PAIWAN

N. TAWU MT.

S. TAWU MT.

YAMI

LAN HSU

N

Fig. 3. Topographic map of Taiwan showing the distribution of the aboriginal tribes.

1962, based mainly on cultural differences such as language, living and marriage customs, etc., and to a lesser extent on biologic characteristics. It has been thought that the groups of Saisiat, Rukai, and Puyuma are closely associated with the groups of Atayal, Paiwan, and Ami, respectively, and that they may have possibly branched off from them. Therefore some investigators consider them subgroups of the latter three groups (Kano, 1955). Because of the varying interests and emphases of individual investigators it is often difficult to make a clear-cut classification of human races or ethnic groups; sometimes a classification may be consistent in some aspects, but not in others.

There are various subgroups in each of the above tribes. These subgroups are so numerous and of so little significance to this study that they are not listed. Those interested in the subgroups may refer to the studies by the Anthropology Department of Empirical Taiwan University (1935) and by Kano (1955).

Two aboriginal groups have been excluded from the list because of the extent to which they have lost their tribal identity. One of these is the so-called Ping Pu tribe. (Ping Pu in Chinese means low land or flat land.) The Ping Pu people once lived on the west side of the island, where they served as a barrier or "buffer zone" between the Chinese and some of the mountain tribes, greatly cutting down contact between them. They were gradually pushed by the Chinese to higher altitudes close to the Central Mountains; some crossed the mountains and settled on the east side.

The Ping Pu are considered a civilized tribe. They tend to mingle with the Chinese, living in the same villages with them. They have practically lost their native culture and identity; not many members of this tribe can speak their original language. The relationships of the Ping Pu tribe to other indigenous tribes and to the early immigrants are not known, but judging by the directness of their contact with later settlers it seems likely that there was some infiltration, genetic as well as cultural, even then. The rate of infiltration has increased with time.

This tribe has many subgroups, and the population during the Dutch occupation was about 8000. The Japanese recorded about 50,000 in 1927 (Chang, 1951). No registration as a separate ethnic group under the Chinese government has been made; so present population figures are unknown. Since the people of this tribe inhabit many different villages their geographic distribution is difficult to define and they are therefore not shown on the map.

The other tribe, relatively less civilized than the Ping Pu but more distinct in a social and cultural sense, is the Shao tribe. It is a small population of about 160, according to recent statistics. Since Shao tribal rules forbid marriage between close relatives, a population of this size can hardly survive without intermarriages with people outside its own tribe. Recent studies of Shao family structure show that the tribe frequently adopts Chinese or other tribe people and intermarriages take place after the adoption (Chen et al., 1955).

Besides the known aboriginal tribes it has been reported (Wei, 1954) that there were once other native groups, now extinct, and that one of these, called "little black man" by the Chinese and various other names by the different aboriginal tribes, was distributed widely on the west side of the Central Mountains. They were described as short, dark-skinned people with short curly hair, who used bows and arrows, were good swimmers, were superstitious about the custom of tattoo, and lived in mountain caves. These people, presumably Negritos, disappeared about 100 years ago. Their existence was mentioned in many Chinese documents of the Ching Dynasty concerning Taiwan (Wei, 1954).

There were also, according to Riess, and a Japanese investigator (Wei, 1954), other groups that became extinct 100 years ago; but they are too genetically remote to be mentioned here.

An arrangement of the aboriginal tribes in order of cultural advancement was made by Chen et al, 1955, as follows:

Civilized ⟵——————————————————⟶ Primitive

Ping Pu	Shao	Saisiat	Tsou	Atayal
	Puyuma	Ami	Paiwan	Bunun
				Rukai
				Yami

The geographic distribution of the existing Taiwan aboriginal tribes is shown on the map (Fig. 3), and the percentages of people living at various altitudes are given in Table 2. The major source of the following statements is the study made by the Anthropology Department of Taiwan Empirical University.

The Atayal tribe inhabits the northern mountain regions. About 50 percent of the population live at an altitude of

Table 2. The distribution of the aboriginal populations in percentages at various altitudes (after Chen (1960) from Kano).

Tribes	Altitudes (feet)					
	<325	325–1625	1625–3250	3250–4875	4875–6500	>6500
Atayal	8.4	18.3	28.4	34.5	10.4	—
Saisiat	—	—	95.9	4.1	—	—
Bunun	—	9.2	22.7	38.2	26.0	3.9
Tsou	—	13.3	34.5	52.2	—	—
Paiwan & Rukai	1.6	32.1	56.6	9.7	—	—
Puyuma	58.6	41.4	—	—	—	—
Ami	56.2	43.8	—	—	—	—
Yami	100.0	—	—	—	—	—
Total	22.9	28.8	25.8	16.7	5.2	0.6

from 3000 to 4500 feet above sea level; the highest altitude at which they live is 5020 feet. There is evidence that they once inhabited the plain region close to the west shore and moved gradually into the mountain area. Some of them crossed over the Central Mountains and settled on the east side.

Many wars have been fought by this tribe, the most recent being the WuShei battle against the Japanese Army, when villages were burned and lives were taken. Events of this sort may have caused much movement of the tribe within its own geographic district.

In close physical contact to the Atayal on the west is the small Saisiat tribe, which inhabits the mountainsides at a relatively lower altitude. Since the Saisiats are bordered on the east by the Atayal and on the west by the native Chinese, they face territorial threats from the Atayal and a cultural threat from both sides.

The Bunun tribe lives in the middle Central Mountains, covering a rather large territory. About 70 percent of the people live at an altitude of from 4000 to 7000 feet; this group's average altitude is the highest among the tribes. It was reported that the Bununs originally populated the plain along the Chu Shui River, then moved along the river to the mountains. Some families crossed the Central Mountains, and some moved to the mountain region farther south about 150 to 200 years ago. The center of the population is considered to be north of Yu Mountain. Although they are surrounded entirely by other tribes, the Bununs are reported to be the least contaminated, both culturally and genetically.

The Tsous live west of Yu Mountain. Like the Atayals and Bununs they are mountain settlers. More than half of the tribe lives at altitudes ranging from 3000 to 4500 feet. Possibly there has been both cultural and genetic infiltration from the neighboring tribes, Bunun in the north, Paiwan in the south, and Ping Pu in the west.

The Paiwans inhabit the most southern part of the Central Mountains, populating densely around Tawu Mountain. It has been reported that the Paiwans originally lived on the plain and in the hilly land close to the west coast, but gradually moved inward to the mountainsides. More than 50 percent of the people live at an altitude between 1500 and 3000 feet. There appears to be some relationship with the Ping Pu tribes remaining on the western outskirts of this area. Some subgroups of the Paiwan tribe crossed the Central Mountains and settled on its eastern slope, and some moved farther down, close to the tip of the island.

The Rukais are distributed in an area west of the Central Mountains. Their neighbors are the Bununs and Tsous in the north, and the Paiwans in the south. As in certain other tribes, some families of the Rukais crossed the Central Mountains and settled permanently on the east side, becoming neighbors of the Puyumas.

A new settlement is growing in an area that originally was too dry for habitation before a water supply was provided a few years ago. Villages in this area are inhabited by Rukais, Paiwans, and retired Chinese military men.

All the above tribes are considered Western tribes, although some subgroups have crossed the Central Mountains and settled on the east side.

The Puyumas and Amis are east-coast inhabitants of Taiwan and for this reason they are considered Eastern tribes. Both groups live near sea level. The Puyumas, a small, rather civilized tribe, live together with native Chinese in many villages around the city of Ping Tung.

The Ami tribe lives on a long narrow strip of land on the east coast north of the Puyuma territory. Since they are geographically in direct contact with the Atayals, there have been constant territorial threats.

The only tribe on Lan Hsu island is called Yami, most primitive of the tribes. They depend for their livelihood mainly on fishing. According to Japanese investigators their language is similar to that of Bataan natives north of the Philippine Islands.

♀♂

CHAPTER 4. POPULATION AND FAMILY STRUCTURE

The total population of the Taiwan aborigines was reported as 68,757 in 1650 (Chow and Hsu, 1960). These statistics presumably came from a Dutch source, as the date fell within the period of Dutch occupation, and statistics of the populations in Taiwan were published by the Dutch in a book called "Koloniaal Archief" between 1644 and 1656 (Chang, 1951). No report on the aboriginal population size published between 1650 and 1905 can be found. From 1906 to 1942 statistics were reported at 5-year intervals by the Japanese Government during its occupation, and lately have been by the Chinese government whose latest (1960) figures for total population of the Taiwan aborigines is 218,098. The census method employed by the Dutch is not known, but the statistics reported by both the Japanese and Chinese governments are based on Household Registration records.

Household Registration is taken on a family basis; a family includes relatives who live together. Barclay (1954) has commented on possible errors in the statistics so obtained, but nevertheless these records are the only sources available. In the tribal villages these records are kept by the village managers and in our field trips were often used as references from which to gather people for tests and examinations. Errors were sometimes noted in the registration, such as recent deaths not yet being recorded, and it is pos-

38

sible that the same may have been true for recent births. However, these two types of omissions would tend to balance each other. It is difficult to believe that any families or individuals would deliberately fail to register; I received the impression that the aborigines are very law-abiding. Furthermore, in all the villages and towns we visited the police as well as a village manager supervised public business. For these reasons I believe the statistics of the total aboriginal population in Taiwan to be fairly reliable.

Statistics collected by the Chinese government are available for the total tribal population, and also for individual tribes, not including those living outside the tribal areas (from unpublished statistics, Department of Civil Affairs, Taiwan, 1961). My figures for each tribe are estimated with reference to these statistics and the geographic distribution of the tribe, plus the number of people who live outside the tribal areas. Errors in the estimates must exist, especially for mixed-tribe villages, and though the number of such villages is small, an error would presumably have a great effect on the statistics for small tribes, such as Saisiat, Puyuma, and to some extent, Rukai; it should have less effect on those for the mountain tribes such as Atayal, Bunun, and Tsou; and least on statistics for the Ami and Paiwan tribes, large populations whose geographic distribution is fairly discrete.

Based on Household Registration records, the Taiwan Provincial Government report published by the Department of Civil Affairs for 1962, gives the estimated 1960 population figures for each tribe as follows:

Ami	87,345	Rukai	5,871
Atayal	49,406	Saisiat	3,394
Bunun	21,440	Tsou	3,223
Paiwan	36,775	Yami	1,957
Puyuma	8,687	Total	218,098

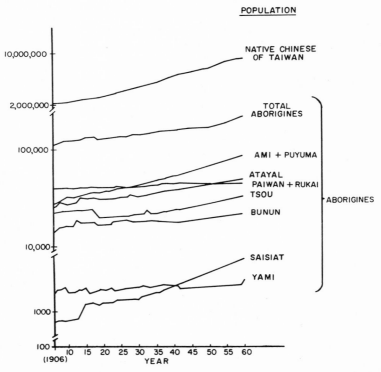

Fig. 4. Population sizes of the different Taiwan aboriginal tribes from 1906 to 1960. Prior to 1945 the statistics were reported during Japanese occupation. After that year they were reported by the Taiwan Provincial Government of China. No individual tribal statistics are available after 1942. The statistics for the individual tribes in 1960 were estimated. The population of the native Chinese in Taiwan is also plotted for comparison.

According to these statistics the total aboriginal population in Taiwan has increased about three-fold in 310 years.

Population growth for the period 1906–1960 in each tribe and in all tribes together (Chen and Divan, 1951) is shown in Fig. 4, with the growth of the native Chinese populations in the same period given for comparison. Among the aboriginal tribes, the Saisiat and Ami grew most rapidly; the Atayal and Tsou were intermediate; and the Bunun, Paiwan, Rukai, and Yami tribes grew the least. Living conditions and physical environment may be important factors affecting population size. The Amis and Saisiats, who live at the relatively lower altitudes, have been changing from a primitive to a civilized type of living, as have the Puyumas, who live mostly in mixed villages with the native Chinese. Competition between the Puyumas and the Chinese may be a factor limiting the growth of the Puyuma population. The mountain tribes, which have shown less cultural advance, are of relatively low population density, but their physical environments are very poor. The primitive Yamis, on the isolated island of Lan Hsu, have a high incidence of malnutrition; the limiting factor to their population increase is physical environment. The Paiwan and Rukai may also suffer from geographic limitation. It is difficult to explain a variation in population size from one year to the next, as seen in many tribes in the past, except as a result of starvation, infectious disease, or war. There may, of course, have been errors in statistics also; but in view of the large numerical fluctuation between seasons or years in ecologic populations (Ford, 1964) it would hardly be reasonable completely to eliminate the possibility that such variations actually occurred.

The age distribution for the whole aboriginal population is illustrated in a diagram of a so-called population pyramid (Fig. 5) based on the statistics published by the Chinese government. In one pyramid the aboriginal population statistics are compared with those of the native Chinese in Taiwan, and in the other with those of the United States (U.S. Bureau of the Census, 1960). The aboriginal population pyramid shows a relatively large base tapering rapidly toward the top, signifying a typical primitive population with

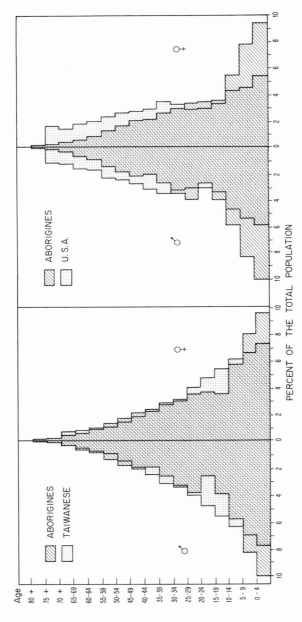

Fig. 5. Population pyramids. On the left of the chart are those for the aborigines and the native Chinese population in Taiwan; on the right, for the aborigines and the population of the United States. Larger mortality is shown in the aborigines than in the Taiwanese at younger ages only. Between the aborigines and the U. S. population greater mortality of the aborigines runs through all ages, but is most pronounced in the youngest and oldest. Note also the large base of the aborigines pyramid, indicating large birth rate.

Table 3. Infant mortality of Taiwan aborigines living in the tribal areas in 1962 (statistics from Taiwan Provincial Government).

Sex	No. born	No. dead	Percent dead
Male	2856	895	31.8
Female	2693	759	28.2
Total	5549	1654	29.8

large birth rate, and high mortality at young and old ages. Infant mortality is extremely high among the aborigines, with an average of 288 per 1000 infants (Table 3) (from statistics of Department of Civil Affairs, Taiwan, 1963). Since some neonatal deaths may not be reported, actual mortality may be higher.

The fertility rate, defined as the number of children between 0 and 5 years of age per 1000 women between 15 and 50 years old, was computed for the entire aboriginal population for the years 1959, 1960, and 1961. Values of 92, 93, and 94, respectively, were obtained. These are predictably high in view of the fact that, in spite of severe infant mortality, survival and growth of the populations have been maintained.

FAMILY STRUCTURE

Monogamy is the rule throughout the Taiwan aboriginal tribes, though marriage customs vary. Marriage by purchase with betrothal presents takes place in the Atayal, Saisiat, Bunun, Rukai, Paiwan, and Puyuma tribes. Marriage by interchange, that is, marriage of the husband's sister to his wife's brother, occurs in the Saisiat, Bunun, and Tsou tribes. Marriage by capture was practiced by the Atayal, Bunun, Paiwan, and Rukai tribes at one time, but has not been reported recently. Choice of spouse is limited by the social structure within the tribes. Marriages in the tribes that have the clan system, such as the Saisiat, Bunun, Tsou, and Ami, are clan exogamous. In those tribes that have the caste system, such as the Rukai and Paiwan, marriages are caste endogamous. The clan system has not been strictly kept in the Puyuma tribe and therefore clan exogamy is not fully followed (Chen, 1955).

Studying consanguinity in any of these tribes is difficult because none of the tribes has a written language and some have complicated naming systems. It appears that the tribes having the clan system may have less inbreeding than those with the caste system. Although as a rule practically all the tribes forbid marriage with relatives closer than second half-cousins, I believe some marriages may have a closer relationship than is actually realized. This is not only because some tribes are rather small, so that choice of spouses of similar age and comparable status is limited, but most people have intermarried in the same village or in closely neighboring villages. For instance, Chen (1955) found that in a Rukai village of 1025 people, 54 marriages had taken place between members of the same village, and 31 with members of neighboring villages. On the other hand, movement within tribal areas from one territory to another because of war, famine, etc., causing the extinction of old villages and abandonment of old neighborhoods, has undoubtedly cut down inbreeding. Such movements occurred often in the history of almost every tribe.

There are two different types of family structure among the Taiwan aborigines: nuclear and composite. A nuclear type comprises a married couple and their immediate descendants, while a composite type may represent a polygamous family, or an extended family which includes two or more married couples of parent-offspring or sib relationships, and their descendants, if any. Since as a rule there are no polygamous marriages among the aborigines, the composite type here refers only to an extended family. Chen

(1955), surveying a village of 189 households (1025 people) in the Rukai tribe, found that 53 percent of the families were of the nuclear type, 39 percent of the composite type, and the remaining 8 percent neither nuclear nor composite, as most of the members were only remotely related. Wang (1959) surveyed the family structure of a Paiwan village of 410 people in 75 households. He reported that 52 percent of the families were of the nuclear type, 43 percent of the composite, and others made up the remaining 5 percent. These estimates are fairly close to those obtained by Chen.

The average number of members of a household varies among the aboriginal tribes (Table 4) (Chen, 1955). The statistics appearing in the table were based on documents published by the government of Taiwan during Japanese occupation. Statistics for the period 1921–1936 show that the Bunun, Ami, and Tsou tribes had large average households, with the number of adult members ranging from 7.5 to 9.3; and the other tribes had comparatively small households, averaging 4.5 to 5.5 members. There is a general decrease in average household size shown in the 1961 statistics (Table 5). These figures are compiled from the Household Registration taken by the Chinese government. They are not average for the entire tribal populations but they include a large number of the members. The decrease

Table 4. Average number of people in a family in the Taiwan aboriginal tribes (after Chen, 1955).

Year	Tribe								
	Atayal	Saisiat	Bunun	Tsou	Rukai	Paiwan	Puyuma	Ami	Yami
1921	4.73	5.87	9.24	7.62	4.49	5.26	5.64	7.76	4.85
1926	4.76	5.49	9.59	8.42	4.53	5.43	5.48	7.97	4.59
1931	4.76	5.45	9.40	7.68	4.77	5.29	5.03	7.76	4.59
1936	4.87	5.76	8.96	6.09	4.86	5.33	5.54	7.97	4.28
Avg.	4.78	5.52	9.29	7.45	4.66	5.33	5.42	7.86	4.58

Table 5. Number of households and people of the different tribes in villages distributed in the tribal areas (1961).

Tribe	No. of households	No. of people	Average number of people in a household
Atayal	1591	7835	4.92
Saisiat	336	2108	6.27
Bunun	951	5891	6.19
Tsou	466	2719	5.83
Paiwan	3985	22310	5.60
Rukai	779	4785	6.14
Puyuma[a]			
Ami	6766	43687	6.46
Yami	473	1957	4.14

[a] Statistics unobtainable.

in family size with time, which occurs particularly in tribes having large families, is presumably due to the military draft, since enlisted men are not included in the Household Registration. In addition, there may have been an actual reduction of family size resulting from social and economic changes after the Chinese government recovered Taiwan in 1945.

Since the number of different relatives contained in a household varies among the tribes, the statistics in Table 4 provide little of the true picture of family size. To obtain the statistics for the number of children per marriage in each tribe would have required a special investigation which was not possible in our survey. Nevertheless, some pedigree data including two to four generations were collected in a large village (Lo-Na) of the Bunun tribe with the assistance of the local school teachers and the village manager. (This village is one of those we visited in our survey. As mentioned in the previous chapter, the Bunun tribe is one of the least civilized, and geographically more isolated than the others on Taiwan.

Table 6. Number of mothers having surviving children in one village of the Bunun tribe, grouped according to age (1963).

No. of children	Mother's age									
	<20	20–24	25–29	30–34	35–39	40–44	45–49	50–54	55–59	>59
0	2	2			1		3	4	2	4
1	2	10	4	3	1	3	11	4	3	14
2		1	4	5	5	2	6	5	3	12
3			6	9	7	6	5	4	6	2
4			8	13	10	7	6	2	1	2
5			2	4	6	6	4	3		2
6				1	4	7	1			
7					2	3	1		1	
8						1	1			
Total no. of mothers	4	13	24	35	36	35	38	22	16	36
Average no. of children per mother	0.5	0.92	3.00	3.37	3.89	4.40	2.71	2.13	2.38	1.72

The people who live in this village have had a stable recent history.) From these data, the number of children produced per marriage were grouped according to the mothers' ages (Table 6). Of a total of 260 marriages for the whole village, 18 had no children and the largest number of children per marriage was 8. The number increases up to the age of 40 to 44 years and decreases thereafter, apparently because of the relative balance between birth and mortality rates. Before the age of 45 to 49, the rate of birth is higher than the rate of death, and from that age on the death rate is higher. For mothers 50 years old and older the youngest children are about 10 years of age, and the number of such offspring left per marriage may be considered the potential breeders for that generation. The total number of children of mothers of 50 years old and over in this village is 149 for 75 marriages, or an average of just about 2 surviving offspring per marriage—enough to maintain the population. Although this was a rather small sample, the picture it provides agrees quite well with the known slow rate of population growth of the Bunun tribe (Fig. 4).

Succession in the matter of property and of family authority among the Taiwan aborigines may be classified into three types, according to Chen (1955): (1) primogeniture (inheritance by the first-born) through male progeny, practiced in the Atayal; or sometimes ultimogeniture (inheritance by the youngest son), in the Bunun, Tsou, Saisiat, and Yami tribes; (2) primogeniture through female progeny, practiced in the Ami and Puyuma tribes; and (3) primogeniture regardless of sex, practiced in the Paiwan tribe only. The Rukai practice the first type, but when there is no heir the heiress may take over, having her husband come to live with her. In the three types of transition of family authority, the first practices patrilocal residence (with the exception of the Yami who practice neolocal), the second matrilocal, and the third both patrilocal and matrilocal residence.

♀♂

CHAPTER 5. ANTHROPOSCOPIC OBSERVATIONS

A few descriptive observations have been made of color, shape, and form on somatologic characters of the head, including angle of eye aperture, shape of eyelid, eye fold, eye color, nasal bridge form, thickness of the mucous lips, forms of ear lobe and ear point, hair form, hair color, and skin color. The classifications used are similar to those made by other investigators (Montagu, 1951; Comas, 1960) and widely adopted in physical anthropology. Since observations of this type are prone to subjective error, they were all made by me throughout the survey for the sake of consistency. It must be kept in mind that these observations are based on judgment, with reference to models and charts for comparison. My score or classification may be different from what another's would have been for the same individual, so that the findings presented in this section may not be fully comparable with those of other investigators; nevertheless the data should be useful for comparisons between sexes and between the tribes.

Features distorted by permanent defects or by disease could not be included for examination, a circumstance that accounts for differences in sample size of various characters for people in the same tribe. No data are included for the facial characters of school children, since so many relative changes take place in them during growth.

THICKNESS OF THE MUCOUS LIPS

The thickness of mucous lips is classified according to Martin (1928), into four types: thin, medium, thick, and

49

everted. In the thin type the mucous membrane is barely visible; in the medium, it is visible and rounder; in the thick type, it is highly visible and somewhat swollen; while in the everted type the lips are voluminous and protrude markedly outward. The distribution of these four types is given in Table 7 for each sex and each tribe.

Table 7. Percentage distribution of people with different thickness of the mucous lips.

Tribe	No.	Sex	Thin	Medium	Thick	Evert
Atayal	104	♂	16.3	76.0	7.7	0
	174	♀	25.9	70.1	3.4	0.6
Ave.			21.1	73.1	5.6	0.3
Saisiat	67	♂	20.9	76.1	3.0	0
	60	♀	18.3	76.7	5.0	0
Ave.			19.6	76.4	4.0	0
Bunun	118	♂	24.6	68.6	6.8	0
	131	♀	17.6	77.1	5.3	0
Ave.			21.1	72.8	6.1	0
Tsou	78	♂	21.8	73.1	5.1	0
	78	♀	23.1	76.9	0	0
Ave.			22.5	75.0	2.6	0
Paiwan	144	♂	31.9	66.0	2.1	0
	240	♀	19.6	76.7	3.3	0.4
Ave.			25.8	71.4	2.7	0.2
Rukai	116	♂	11.2	85.3	3.4	0
	102	♀	8.8	86.3	4.9	0
Ave.			10.0	85.8	4.2	0
Puyuma	73	♂	16.4	69.9	13.7	0
	118	♀	3.4	85.6	10.2	0.8
Ave.			9.9	77.6	12.0	0.4
Ami	163	♂	6.7	85.3	8.0	0
	94	♀	7.4	83.0	9.6	0
Ave.			7.1	84.2	8.8	0

In all the tribal groups the majority of people are of the medium type, ranging from 66.0 percent for the men of Paiwan to 86.3 percent for the women of Rukai. The remaining population is distributed more in the thin class than in the thick. Only three women of three different tribes were classified in the everted type.

Among the tribes there is a tendency toward a higher frequency of thicker lips in the Ami and the Puyuma tribes, which differ significantly from each of the other tribes at the 5 percent level by the chi-square test. (Throughout this monograph, unless otherwise indicated, less than 5 percent probability is considered significant.)

Sex difference in lip size varies among the tribes, and the variation is greater in the thin lip class than in the others. Differences in lip size between sexes are small in the Saisiat, Tsou, Rukai, and Ami tribes. In the remaining tribes sex differences appear large but vary in direction and are not consistent with respect to the increasing order of classification. Although by the chi-square test some of the differences are significant, the class which contributes about 80 percent or more to the chi square is the thin class, in which the number of people so classified is rather small in comparison with that of the medium type in each tribe. These differences are practically nonexistent when all the tribes are pooled together. Furthermore, the fullness of the lips is partially affected by age and health, so that the lips of older or undernourished persons would appear thinner than those of the young and healthy. It is very doubtful whether there is any real genetic difference in lip thickness between the sexes.

NASAL BRIDGE

The shape of the nasal bridge is classified into three types according to the criterion of Martin as illustrated in Comas (1960): concave, straight, and convex. The results of our

Table 8. Percentage distribution of people with different nasal bridge forms.

Tribe	No.	Sex	Nasal bridge		
			Concave	Straight	Convex
Atayal	107	♂	72.8	27.1	0
	172	♀	90.1	9.8	0
Ave.			81.5	18.5	0
Saisiat	68	♂	51.4	45.6	2.9
	59	♀	69.5	28.8	1.7
Ave.			60.5	37.2	2.3
Bunun	119	♂	68.9	26.9	4.2
	130	♀	83.1	16.2	0.8
Ave.			70.7	21.6	2.5
Tsou	81	♂	30.9	65.4	3.7
	78	♀	48.7	47.4	3.8
Ave.			39.8	55.9	3.8
Paiwan	143	♂	49.0	49.6	1.4
	239	♀	77.4	22.6	0
Ave.			63.2	36.1	0.7
Rukai	119	♂	42.0	57.1	0.8
	101	♀	74.3	25.7	0
Ave.			58.2	39.4	0.6
Puyuma	71	♂	47.9	50.7	1.4
	116	♀	81.9	17.2	0.9
Ave.			64.9	34.0	1.2
Ami	151	♂	50.3	47.7	2.0
	103	♀	65.0	34.0	1.0
Ave.			52.7	40.9	1.5

observations are given in Table 8 for each sex in each of the eight tribes. Most of the subjects were either of the concave or the straight type; people with convex nose bridges were very rare, ranging from 0 percent in Atayal to 4.2 percent in Bunun males. Between the other two types there were more concave than straight in each tribe. A difference between sexes was noted in all the tribes, with percentages of concave nose type greater and of straight nose type smaller in women than in men.

Since the convex nose type was extremely rare, the nose bridge form became a character of dichotomy for the tribal populations, an increase in the percentage of one category being at the expense of the other. For this reason, if we wish to compare the differences between tribes, we can examine only a single classification separately for men and women. For the concave type the tribes are arranged in decreasing order for men: Atayal, Bunun, Saisiat, Ami, Paiwan, Puyuma, Rukai, and Tsou; and for women: Atayal, Bunun, Puyuma, Paiwan, Rukai, Saisiat, Ami, and Tsou. There are some interchanges in rank order among the tribes between highest and lowest, but the first and second and the last tribes are consistent between sexes. Differences in percentage between two tribes smaller than 15 percent are in general not significant.

EYE APERTURE

The eyelid opening is horizontal in most individuals of the human species, but there are some in whom it is inclined, so that either the inner canthus is lower than the outer, or vice versa, causing the eye opening to appear oblique. I do not think this inclination is caused by the existence of a skin fold of the upper eyelid which covers the free margin of the eyelid, although this has been so claimed (Comas, 1960, p. 271); this will be discussed further in connection with the Mongolian fold.

The distribution among the three types of eye aperture (horizontal, inclination inward-down, and inclination outward-down) is given in Table 9. There is no significant difference between left and right eyes, or between sexes.

Table 9. Percentage distribution of people with different eye apertures.

| Tribe | No. | Eye aperture | | |
		Horizontal	Inward-down	Outward-down
Atayal	281	96.8	2.8	0.4
Saisiat	127	92.9	4.7	2.4
Bunun	249	96.8	1.2	2.0
Tsou	159	80.5	3.1	16.4
Paiwan	162	91.4	5.2	3.4
Rukai	219	97.3	1.4	1.4
Puyuma	189	97.4	2.1	0.5
Ami	253	96.0	2.4	1.6

Therefore the statistics for right and left eyes, and for men and women, are pooled together for each tribe.

It will be noticed in Table 9 that the majority of eye apertures are in the horizontal category. However, there is a relatively high percentage of Tsou people in the category of inclination outward-down, and a relatively high percentage of Paiwan people in the inclination inward-down category, and the differences between each of these two and the remaining tribes is significant. But the actual numbers of people in the rare categories are small in comparison with those in the horizontal. For this reason we need to be cautious in drawing definite conclusions from the differences.

EYEFOLD

"Eyefold" refers to the arrangement of the skin over the upper eyelid and canthi. When the skin above the upper eyelid is loose and hangs down over the free margin of the eyelid, it is called complete Mongolian fold. Skin that hangs over merely the inner canthus is called the internal epicanthic fold, that over the middle part is the median fold, and that over the outer canthus is the external epicanthic fold.

Skin that is tight and forms an arch above the upper eyelid is called no-fold. These are the general classifications of eyefolds given in the textbook by Montagu (1951) from Martin (1928).

In this study there are two general classifications of eyefold: Mongolian and internal epicanthic. The Mongolian folds are further classified into internal, median, external, complete, and none; and the epicanthic folds, concerned only with the inner canthus, are further classified into none, slight, and extensive.

During the course of the examinations it was found that many people who appear to be of the no-fold type have a slight skin fold growing out between the ordinary skin arch and the free margin of the eyelid in the inner canthic region which forms a fold to *cover,* and circle around, the inner canthus (Fig. 6). The extent to which it extends below the inner canthus varies. It was also observed that many individuals with either the complete or internal epicanthic fold of the type described by Montagu were without the above skin fold.

The difference between this and the Mongolian fold was recognized by Aichel in 1932 according to Gates (1946), and Whitney (1942). Gates rightly pointed out that Chouke (1929) mistakenly suggested that the terms epicanthus and Mongolian fold should be used interchangeably. The distinction between epicanthus and Mongolian fold lies in the fact that the epicanthus is extended vertically across the inner or outer canthus, whereas the Mongolian fold extends laterally to cover part or all of the free margin of the eyelid. Mongolian folds are rather loose and less visible in closed eyes, whereas the epicanthi are thin, rather tightly attached, and clearly visible in closed eyes (Fig. 6). We did not check for outer epicanthus *(epicanthus lateralis).* It is my opinion that the inner and outer epicanthi are quite differ-

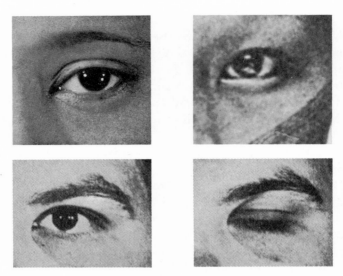

Fig. 6. (*Top left*) The left eye of a person with the epicanthus without Mongolian fold (after Krauss, 1942). (*Top right*) A slight epicanthus and complete Mongolian fold of an Atayal woman; the black pattern on the face was made by tattooing, which is practiced in this tribe. (*Bottom*) The epicanthus of a German woman, eyes open and closed. (From R. R. Gates (after Aichel), *Human Genetics,* 1946. Courtesy of the MacMillan Company, New York.)

ent in morphology and possibly in genetics, but that definite racial characteristics relating to the outer epicanthus have not been reported.

At the time we discovered the distinction between Mongolian and internal epicanthic fold our field survey was already half over. Therefore, data concerning the epicanthic fold cover only four tribes. Distribution into the different epicanthic eyefold classes of each sex of each of the four tribes is given in Table 10. The larger proportion of aborigines, ranging from 61 percent of the Ami women to 85 percent of the Rukai men, have no epicanthic fold.

Table 10.   Percentage distribution of people with or without epicanthus.

Tribe	No.	Sex	Epicanthus		
			None	Slight	Extensive
Paiwan	42	♂	79.2	17.0	3.8
	50	♀	79.4	17.5	3.2
Ave.			79.3	17.3	3.5
Rukai	51	♂	85.0	15.0	0
	43	♀	84.3	13.7	2.0
Ave.			84.7	14.4	1.0
Puyuma	60	♂	84.5	14.0	1.4
	80	♀	71.4	21.4	7.1
Ave.			78.0	17.7	4.3
Ami	119	♂	77.8	20.9	1.3
	55	♀	61.1	33.3	5.6
Ave.			69.5	27.1	3.5

When epicanthic folds are found, there appears to be a difference between sexes, with relatively more women having them than men, and these differences in the Puyuma and Ami tribes are significant. The epicanthus is claimed to behave as a single recessive Mendelian character (Fortuyn, 1932). I am not convinced that this is the case, however, for there is larger variation in its visibility and length. Sometimes when least visible it is difficult to distinguish from nonepicanthus. The epicanthus appears to be more characteristic of the Mongolian race and its derivatives than the Mongolian fold.

Distribution of the different types of Mongolian folds is given in Table 11 for each sex of each tribe. The number of people having the no-fold type ranges from 51 percent for Atayal women to 96 percent for Tsou women. Sex differences appear, but they are small in most tribes and vary in direction; there is probably no real sex difference. For

Table 11.   Percentage distribution of people with different Mongolian folds.

Tribe	No.	Sex	Mongolian fold				
			Complete	Internal	None	External	Median
Atayal	214	♂	15.7	5.1	64.0	15.0	0
	348	♀	21.3	10.8	50.9	16.4	1.2
Ave.			18.5	7.7	57.5	15.7	0.6
Saisiat	134	♂	3.0	0	94.0	3.0	0
	122	♀	5.7	0.8	86.8	6.6	0
Ave.			4.4	0.4	90.4	4.8	0
Bunun	258	♂	3.9	0	81.5	13.9	0.8
	260	♀	4.2	4.6	74.1	16.1	0.8
Ave.			4.1	2.3	77.8	15.0	0.8
Tsou	154	♂	1.3	0.7	91.5	6.5	0
	162	♀	0.6	1.9	96.3	1.2	0
Ave.			1.0	1.3	93.9	3.8	0
Paiwan	288	♂	3.1	0	84.7	12.1	0
	479	♀	2.3	0.2	88.6	8.5	0
Ave.			2.7	0.1	86.7	10.3	0
Rukai	232	♂	6.0	0	88.8	5.2	0
	202	♀	3.9	0	92.1	4.0	0
Ave.			5.0	0	90.5	4.6	0
Puyuma	146	♂	4.1	0	88.9	6.8	0
	236	♀	2.1	0.8	91.4	5.5	0
Ave.			3.1	0.4	90.2	6.2	0
Ami	326	♂	2.8	0	89.4	7.3	0.3
	188	♀	3.2	0	74.9	21.8	0
Ave.			3.0	0	82.2	14.6	0.2

comparison between tribes, we pooled the sexes. More people with inner and outer folds, and fewer with no folds were found in the Atayal than in the other tribes, the differences in the percentages of eyefold between this tribe and the others being significant in most cases.

IRIS COLOR

The subjects were examined for iris color close to a window or an open door, with indirect sunlight as the light source. The iris color was matched to a color chart manufactured by J. F. Lehmans Verlag, Munich, based on Martin and Schultz's classification, which includes different shades of color ranging from light blues and browns to black brown. This chart was not available at the beginning of our field survey, so that examinations for iris color were not made for people of the Saisiat nor for some individuals of the Atayal tribe. The distribution of iris color in each sex in each tribe is shown in Table 12. Sex differences appear and vary among the tribes; the general tendency is toward darker iris color in women than in men. Among the tribes, people of the east coast tribes (Ami and Puyuma) appear to have lighter iris color and the Bunun a comparatively large proportion of dark iris color.

EAR LOBE AND EAR POINT

The form of the ear lobe is classified into two types, adherent and free. In the free type it varies from a small lobe, with the lower edge slightly below the point of the attachment of the ear, to a rather large lobe protruding downward. This variation appears to be quantitative rather than qualitative. Since it is difficult to make clear-cut separations according to size and protrusion of the point, no separation was made in the free-lobe class. In general the free type is ordinarily easier to separate from the adherent type, but the difference again is quantitative, and the character is no more a dichotomous one. Differences appear between sexes within the tribes, but they vary in direction (Table 13), and are possibly sampling or observation errors. There are also differences between tribes.

Table 12. Percentage distribution of people with different iris color.

Tribe	No.	Sex	Iris color											
			1B	1C	2	2A	2B	2C	3	4	4A	4B	5	6
Atayal	77	♂	0	0	0	0	0	0	0	0	0	6.4	63.2	29.7
	114	♀	0	0	0	0	0	0	0	0.9	0.9	10.3	55.0	31.0
Ave.			0	0	0	0	0	0	0	0.5	0.5	8.4	59.1	30.4
Bunun	124	♂	0	0	0	0	0	0	0	0.8	0	0.8	60.8	36.8
	131	♀	0	0	0	0	0	0	0	0.8	0	0	40.3	58.5
Ave.			0	0	0	0	0	0	0	0.8	0	0.4	50.6	47.7
Tsou	80	♂	0	0	0	0	0	0	0	0	6.2	15.0	42.5	36.2
	81	♀	0	0	0	0	1.2	0	1.2	0	6.2	11.1	33.3	46.9
Ave.			0	0	0	0	0.6	0	0.6	0	6.2	13.1	37.9	41.6
Paiwan	144	♂	0	0	0	0.7	2.8	0	4.2	0	15.3	26.4	30.6	20.1
	240	♀	0	0	0.4	0	0	0	4.6	0	10.8	21.2	32.9	30.0
Ave.			0	0	0.2	0.4	1.4	0	4.4	0	13.1	23.8	31.8	25.1
Rukai	116	♂	0	0	0	0	0	0	3.4	0	20.7	24.1	37.1	14.6
	100	♀	0	0	0	0	0	0	2.0	0	17.0	19.0	30.0	32.0
Ave.			0	0	0	0	0	0	2.7	0	18.9	21.6	33.6	23.3
Puyuma	73	♂	0	0	0	0	0	0	11.0	0	21.9	31.5	26.0	9.6
	118	♀	0	0	0	0	0.8	0	3.4	0	15.2	28.8	36.4	15.2
Ave.			0	0	0	0	0.4	0	7.2	0	18.6	30.2	31.2	12.4
Ami	161	♂	0.6	0.6	0	0	6.8	0.6	17.4	0	16.8	31.7	19.2	6.2
	94	♀	0	0	0	0	2.1	0	11.7	0	21.3	33.0	27.6	4.2
Ave.			0.3	0.3	0	0	4.5	0.3	14.6	0	19.1	32.4	23.4	5.2

Table 13.   Percentage distribution of people with different types of ear lobe.

Tribe	No.	Sex	Ear lobe	
			Adherent	Free
Atayal	107	♂	40.2	59.8
	174	♀	48.3	51.7
Ave.			44.3	55.7
Saisiat	68	♂	45.6	54.4
	59	♀	42.4	57.7
Ave.			44.0	56.0
Bunun	120	♂	12.5	87.5
	129	♀	39.5	60.4
Ave.			26.0	74.0
Tsou	81	♂	22.2	77.8
	78	♀	34.6	65.4
Ave.			28.4	71.6
Paiwan	145	♂	20.7	79.3
	237	♀	21.1	78.9
Ave.			20.9	79.1
Rukai	118	♂	19.5	80.5
	101	♀	21.8	78.2
Ave.			20.7	79.3
Puyuma	71	♂	28.2	71.8
	118	♀	16.9	83.0
Ave.			22.6	77.4
Ami	150	♂	26.7	73.3
	103	♀	24.3	75.7
Ave.			25.5	74.5

Examination of percentages in some tribes in whom the sex differences are small, such as the Atayal, Saisiat, Paiwan, and Rukai, showed about twice as many of the adherent type of ear lobe in the Atayal and Saisiat as in the Paiwan and Rukai tribes.

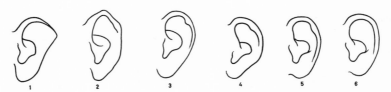

Fig. 7. Types of external ears: 1, macaque form; 2, *cercopithecinae* form; 3, Darwinian point; 4, Darwinian tubercle; 5, vestigial Darwinian tubercle; 6, without Darwinian tubercle. (From J. Comas (after Schwabe), *Manual of Physical Anthropology,* 1960. Courtesy of Charles C Thomas, Publisher, Springfield, Illinois.)

The ear point is classified according to the development of Darwin's tubercle. Schwalbe (Comas, 1960) classified the ear forms into six types: (1) macaque form; (2) ceropithecus form; (3) Darwinian point; (4) Darwinian tubercle; (5) vestigial Darwinian tubercle; and (6) without Darwinian tubercle (Fig. 7). Differences in ear point are relative and quantitative. Sometimes it is rather difficult to assign a person to a class, especially in the No. 4 and 5 types. Therefore in the statistical analysis we combined Schwalbe's No. 4 and 5 classes.

There are quite noticeable sex differences with regard to ear point classification (Table 14). Men have longer ear points than women. Unlike the sex variations in ear lobes, those for ear point are consistent throughout the tribes, and in many of the tribes there are four- or fivefold differences in certain classes. The differences are all significant, and we may accept as fact that they exist between sexes. Between the tribes, Atayals, and Saisiats have the greater proportion of large ear points. Except between Atayal and Tsou, Tsou and Ami, and between each two tribes of Rukai, Puyuma, and Ami, differences between any of the other tribes are significant.

Since a relatively large proportion of Atayals and Saisiats have large ear points and adherent ear lobes, there may be an association between ear-point and ear-lobe type. Such an association might be expected from the evolutionary point of view, since progression from animal to human type involves reduction in ear point and increase in ear lobe. A contingency chi-square test of the association

Table 14. Percentage distribution of people with different ear point (see text for the numerical designations).

Tribe	No.	Sex	Ear point					
			1	2	3	4	5	6
Atayal	107	♂	0.9	2.8	3.7	33.6	16.8	42.0
	173	♀	0	0	1.2	14.4	12.1	72.2
Ave.			0.5	1.4	2.5	24.0	14.5	57.1
Saisiat	67	♂	0	3.0	9.0	44.8	6.0	37.3
	60	♀	0	0	3.3	21.7	13.3	61.7
Ave.			0	1.5	6.2	23.3	9.7	49.5
Bunun	119	♂	0	0	0.8	12.6	1.7	84.9
	131	♀	0	0	0	3.0	0	96.9
Ave.			0	0	0.4	7.8	0.9	90.9
Tsou	80	♂	0	0	2.5	18.8	31.2	47.5
	81	♀	0	0	3.7	3.7	12.3	80.2
Ave.			0	0	3.1	11.3	21.8	63.9
Paiwan	144	♂	0	0.7	2.8	27.1	32.6	36.8
	241	♀	0	0	0.8	14.1	26.1	58.9
Ave.			0	0.4	1.8	20.6	29.4	47.9
Rukai	116	♂	0	0	0	13.8	29.3	56.9
	102	♀	0	0	0	4.9	5.9	89.2
Ave.			0	0	0	9.4	17.6	73.2
Puyuma	73	♂	0	0	0	28.8	31.5	39.7
	119	♀	0	0	0	3.4	17.6	79.0
Ave.			0	0	0	16.1	29.6	59.4
Ami	163	♂	0	0	1.8	14.1	25.2	58.9
	104	♀	0	0	0	11.5	11.5	76.9
Ave.			0	0	0.9	12.8	18.4	67.9

was made for the Atayal and Saisiat data separately. Both the chi-square values were between the 10 and 20 percent probability level. We may conclude that if there is a tendency for protruding point to go with adherent lobe in the human ear it is probably rather slight.

HAIR COLOR AND FORM

Throughout the tribes that we surveyed hair colors were either pure black or black-brown, and the hair form straight and coarse. No light-colored and curved or waved hair was seen. No hair differences between sexes and tribes were observed, except that women's hair is finer than men's.

SKIN COLOR

Because we lacked equipment and standards, skin color was only crudely examined. The skin color of the Taiwan aborigines ranges from pale yellow, as in the Chinese, to chocolate brown as in Malayans. Generally the skin color of the Bununs, Paiwans, and Rukais is darker than that of the other aborigines. During our survey of the Bunun tribe we examined a man with unusually dark skin color, comparable to that of the average American Negro. The color of his wife's skin was typical of that tribe. Their two children's skin color was darker than most of the people of their tribe. This man lived in the same environment as other members of the Bunun tribe; so it is hard to believe this condition is nongenetic. It is likely to have resulted from segregation of genes affecting skin colors. This person is possibly an unusual recombinant of a large number of genes for increasing pigmentation.

Photographs were taken of individuals randomly chosen from each tribe. These are given in Figs. 8–17 as an aid to the descriptions of the somatologic observations. Brief descriptions are given.

Fig. 8. A married Atayal couple.

Fig. 9. (*Left*) Two Saisiat brothers just returned from hunting. In comparison with the other aborigines, both the Atayals and Saisiats have longer noses and faces, and the skin color is lighter than the Bununs and Paiwans, comparable to the Chinese. (*Right*) A Saisiat with the adherent type of ear lobe that also appears in the man in Fig. 8.

Fig. 10. A Bunun woman. Notice her broad nose and round face which is common in the Bununs. She has the free type of ear lobe and is without the Mongolian eyefold. The Bununs are darkest in skin color among the tribes, with coarse and very dark hair.

Fig. 11. Members of the Tsou tribe. The general appearance of the people is somewhat similar to that of the Atayals and Saisiats but they show more prominent nose and cheek bones. The top pictures are of a chieftain of the Tsou tribe.

Fig. 12. Members of the Paiwan tribe. Their general facial appearance is like the Bununs with fairly dark skin color, concave nose, and short stature. Notice the very coarse and dark hair in both persons and the complete Mongolian eyefold in the boy in the lower pictures. I also noticed in many of them rather broad palms and feet in proportion to the length, and short but big fingers.

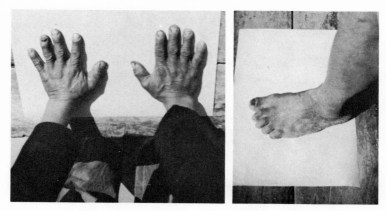

Fig. 13.   The hands and a foot of the man at the top of Fig. 12.

Fig. 14. (*Top*) A Rukai man. Note the round face and concave nose. The Rukais are rather similar to the Paiwans but their skin color is lighter. (*Bottom*) A chieftain of the Puyuma tribe of a more civilized type. He has a fairly light skin color. Larger bone structure of the people of this tribe than people of the other tribes was noticed. The man is balding, but balding is rarely seen in the aborigines.

Fig. 15. Members of the Ami tribe. Like the Puyumas, they are rather civilized. Note the flat and fairly wide distance between the two canthi, light skin color, and fairly fine hair as shown in the woman, traits that are quite common in this tribe.

Fig. 16. Members of the Yami tribe, one of the least civilized, which the author was unable to visit. The pictures were taken by the school principal who gave the intelligence tests. The people of this tribe depend mainly on fishing. They ordinarily wear no clothes.

Fig. 17. (*Top*) The daughter of a Chinese father and a Puyuma mother. Note the close similarities to the Chinese elongated face and nose, flat seated eyes, fairly long forehead, light skin color, and fine hair. (*Bottom*) The daughter of a Japanese father and a Puyuma mother. Note the close similarities to the Japanese: round face, fairly broad and concave nose, slightly slanted eyes, and complete Mongolian folds and slight epicanthi.

DISCUSSION

In the facial characters of human beings, resemblances can be seen between parents and offspring and even between grandparents and grandchildren. This fact, although arrived at entirely by observation, suggests not only a genetic basis for such resemblances but also that the genetics may be comparatively simple. Yet there has been no report covering a full exploration of the genetics of normal variation in facial characters.

As Boyd (1950) pointed out, genetic study of human facial characters may be a matter of finding appropriate methods of coding and recording. This indeed may be the case, and extremely difficult, especially for characters that vary between sexes and between ages. For instance, in the nasal bridge types certain persons can easily be assigned to the right category of straight, concave, or convex type. But for others whose noses are close to the borderline between two types, classification becomes difficult because the difference between one category and another is quantitative, and actually the same subject may be assigned to one category by one examiner and to another by a different examiner. Even the same examiner may assign him to two different categories at two different times. Indeed, as far as normal genetic characters in the human species are concerned, outside of blood groups, serum proteins, and hemoglobin types and some other biochemical traits, none is an all-or-none type. Until more efficient scales or coding methods are developed for the measurement of facial characters, the present classifications, although inadequate, are the best available. Based on these classifications we have found differences between sexes and between the aboriginal tribes. Differences between the tribes may be attributed to both genetic and environmental effects, but the relative importance of the genetic and environmental components cannot be evaluated from the present data.

The comparatively darker skin color of the Bunun aborigines, plus the fact that a very dark-colored family was discovered in the same tribe, raises a rather interesting question of causation. The darker skin color of the Bununs is not likely to be environmental. Other tribes living in the southern part of the island are exposed to about the same amount of sunlight as the Bununs, but have lighter skin color. That the genetics of human skin color depends on multiple factors is quite well established. The frequency of genes for dark skin color must be greater in the Bununs than in the other tribal people. Change of gene frequency can be brought about by mutation, selection and random drift, and migration, and these events could equally likely occur in other tribes. A coincidence in connection with migration concerns a rumor, which we heard from the local people during our visit to the Saisiat tribe, that ". . . back many years ago there used to be a population of small-body sized people with very dark skin color living in the mountain area. People usually called them 'little black man.' They were good-natured fellows, good swimmers and liked to play jokes. . . . These people disappeared not too long ago." This story does not prove that the Bununs contain some genetic elements of the "little black man" but neither can we deny this possibility. Indeed its existence was claimed as mentioned in Chapter 2. Further support of the possibility of genetic infiltration of the Bununs by these people (possibly extending even to other tribes) is found in their stature; the Bununs are the shortest people among the tribes. The genetics of human stature is not quite so well understood as that of skin color but it is believed, from studies in other animal species, that human stature is determined by polygenes or multiple factors greater in number than those for skin color. If the Bununs originally descended from the same stock as the other tribes, changes in stature and skin color must require the change of both sets of gene frequencies. Such an event would more likely be caused by migration than by mutation or random drift. For these reasons I would favor the explanation that the dark skin color and short stature of the Bununs, and to some extent of the Paiwans, are due to infiltration, possibly by the so-called "little black man," the Negrito.

SUMMARY AND CONCLUSION

The observations on somatologic characters in the Taiwan aborigines show that a relatively large proportion of people have medium thick mucous lips, a concave nasal bridge, and iris of a dark brown color. Although there are people with slanted eye apertures, the majority have the horizontal type; similarly, many people have the Mongolian fold—but a large proportion are without. Most of the people have no epicanthic fold. Although a few individuals in certain tribes have large ear points, the majority have the common type of external ear. Skin color ranges from pale yellow to chocolate, and hair is either black or black-brown, practically all straight and fairly coarse. Differences between sexes and between tribes appear in many of the anthroposcopic characters examined. Unquestionably the aborigines contain more of Mongoloid than of any other basic elements.

Characters that differ most consistently between men and women are the nasal bridge, the epicanthus, iris color, ear point, and hair type. Relatively more women have concave nasal bridges, more epicanthi, darker iris color, smaller ear points, and finer hair.

Among the tribes, relatively more Ami and Puyuma people have thick mucous lips, and lighter iris and skin color. Large ear points and adherent ear types are characteristic of the Atayals and Saisiats. In the Atayal and Bunun tribes a comparatively large percentage of people are of the concave nasal type. The straight type of nasal bridge is found more often among the Tsous than elsewhere. Characteristic of the Bununs is their darker skin color and color of iris. A comparatively large percentage of Tsou and Paiwan people have slanted eye apertures, although the horizontal type is still predominant in these tribes. The Ami people have the most epicanthi. These findings are incorporated for further discussion in the anthropometric measurements given in the next chapter, in which tribal resemblances are more fully discussed.

♀♂

CHAPTER 6. ANTHROPOMETRIC MEASUREMENTS

In making head and body measurements we selected dimensions that were of relatively high importance in racial classification and easy to obtain accurately, and rejected those highly correlated with measurements already taken or lacking in clear-cut landmarks. Among the latter were acrominal height, pubic arc height, bispinal diameter, and body weight, which were measured but omitted from the analysis. A total of 11 head and 8 body measurements on each adult aborigine were used in the following analyses.

LINEAR MEASUREMENTS

So that this monograph may be self-contained for readers not familiar with anthropometric measurements, the landmarks for both head and body measurements are illustrated (Figs. 18 and 19) and brief descriptions of procedures given (Comas, 1960).

Head Measurements

Head length. The maximum glabella-occipital length in the median sagittal plane was measured by placing one point of the spreading calipers on the glabella and moving the other up and down the median occipital line until a maximum reading was obtained.

Head breadth. The maximum transverse diameter of the vault was obtained by keeping the points of the calipers in one horizontal and lateral verticular plane and moving them forward and backward and up and down until a maximum reading was reached.

Head height. The distance from tragion to bregma was obtained by subtracting the tragion height from the stature. Direct measurement, which would have been more accurate,

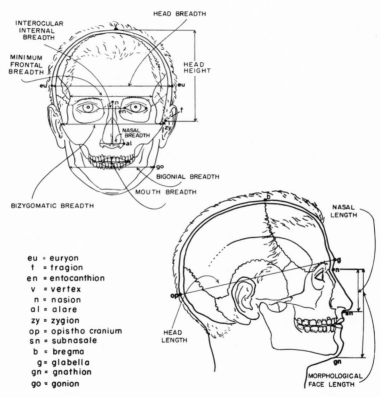

eu = euryon
t = tragion
en = entocanthion
v = vertex
n = nasion
al = alare
zy = zygion
op = opistho cranium
sn = subnasale
b = bregma
g = glabella
gn = gnathion
go = gonion

Fig. 18. The landmarks on the head used for taking the anthropometric measurements: *eu*, euryon; *t*, tragion; *en*, entocanthion; *v*, vertex; *n*, nasion; *al*, alae; *zy*, zygion; *op*, opisthocranium; *sn*, subnasale; *b*, bregma; *g*, glabella; *gn*, gnathion; *go*, gonion. (From J. Comas, *Manual of Physical Anthropology,* 1960. Courtesy of Charles C Thomas, Publisher, Springfield, Illinois).

could not be used because of the subjects' objections to insertion of the caliper point in the ear.

Minimum frontal breadth. The minimum diameter between temporal crests was taken by determining the temporal crests with thumb and forefinger and then applying the points of the calipers.

Bizygomatic breadth. The maximal diameter between the zygomatic arches was taken by holding the arms of the calipers while palpating the zygomatic arches to make sure the points of the calipers were correctly applied.

Bigonial breadth. The diameter between external surfaces of the gonia was obtained by placing the points of the calipers on the most laterally projecting angles of the mandible at the junction of the ascending with the horizontal rami.

Interocular internal breadth. The transverse distance between the inner canthi was taken with the sliding calipers, the points barely touching the skin at the inner canthi.

Morphologic facial length. The distance from nasion to gnathion was measured with the subject sitting, mouth closed. The nasion was located by palpation to find the groove marking the nasofrontal suture, and the points of the sliding calipers were then applied.

Nasal length. The distance from nasion to subnasale was measured with the sliding calipers after locating the nasion as above. The lower point was the junction of the nasal septum with the lip.

Nasal breadth. The maximal diameter of the nasal alae was measured by barely touching the caliper tips to the most laterally situated points on the wings of the nose, with no pressure applied.

Mouth breadth. The maximal diameter between the angles of the mouth at the junctions of mucous membrane and skin was measured with the subject's mouth closed and the tips of the calipers just touching the junction points.

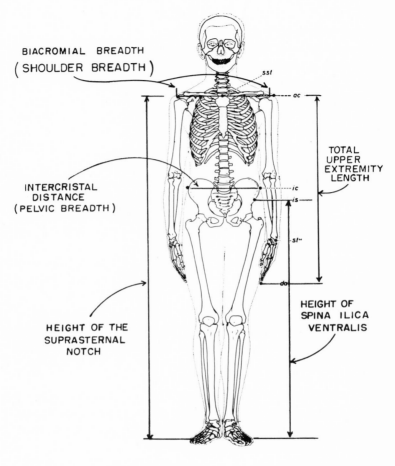

BIACROMIAL BREADTH
(SHOULDER BREADTH)

INTERCRISTAL
DISTANCE
(PELVIC BREADTH)

HEIGHT OF THE
SUPRASTERNAL
NOTCH

TOTAL
UPPER
EXTREMITY
LENGTH

HEIGHT OF
SPINA ILICA
VENTRALIS

sst = suprasternal   ac = acromion
ic = iliocristale   is = iliospinale
da = dactylion

Fig. 19. The landmarks on the body used for taking the anthropometric measurements: *sst*, suprasternal; *ic*, iliocristale; *ac*, acromion; *is*, iliospinale; *da*, dactylion. (From J. Comas, *Manual of Physical Anthropology*, 1960. Courtesy of Charles C Thomas, Publisher, Springfield, Illinois).

Body Measurements

Stature. The subject stood barefoot on a hard level floor and stretched to maximum height with the upper limbs pendant, palms of the hands turned inward. The heels were together, the axis of vision was horizontally forward. Measurement was taken from the vertex with the anthropometer reading to the nearest millimeter.

Suprasternal notch height. The distance from the middle of the anterior-superior border of the manubrium sterni to the floor was measured with the subject in the same standing position as above. The point of the anthropometer was applied to the suprasternal notch in its deepest part.

Spina ilica height. The distance from the summit of the anterior-superior spine of the ilium to the base of the ensiform cartilage sometimes could only be located by tracing Poupart's ligament to its ilica termination which defines the point.

Sitting height. The height of the vertex above the surface of the seat was determined with the subject on a horizontal seat 30 to 40 cm high in the most erect position, knees flexed, gluteal and extensor muscles of the thighs relaxed, and the axis of vision horizontal.

Biacromial diameter. The distance between the two acromial points was measured with the subject in standing position as for stature. The acromial points were located by palpating, and the distance between them was measured with a spreading caliper.

Intercristal distance. The maximum distance between the external margins of the iliac crests was measured with the spreading calipers. The subject was in the same standing position as mentioned under stature.

Chest girth. The maximum contact circumference at the level of the base of the ensiform cartilage was obtained with a metric tape, with the subject in standing position, maintaining natural respiration. Accurate measurement is difficult to obtain, especially in women because of variation in size of the breasts. (To take the extremes of inspiration and expiration is very difficult in the case of aborigines, and we did not attempt it.)

Upper limb length. The distance from the acromion to the dactylion was measured while the subject stood erect with the upper limbs pendant and straight. The measurement was taken with a metric tape between the acromion point and the dactylion.

Symbols representing the various head and body measurements: *Head measurements:* HL, head length; HB, head breadth; HH, head height; FB, minimum frontal breadth; ZB, Bizygomatic breadth; BB, bigonial breadth; OB, interocular breadth; FL, morphologic facial length; NL, nasal length; NB, nasal breadth; MB, mouth breadth; *Body measurements;* ST, stature; SN, suprasternal notch height; IH, spina ilica height; SH, sitting height; BD, biacromial diameter; CD, intercristal distance; CG, chest girth; UL, upper limb length.

The means and standard deviations (s.d.) of the head and body measurements for men and women of each tribe are given in Tables 15 and 16, respectively. In head length the Ami ranks highest, and the Paiwan, Rukai, and Bunun lowest. In head breadth ranks for the first three are reversed. When plotted on a two-dimensional diagram based on the tribe means in terms of standard deviation units (obtained by dividing each mean by the average standard deviation for all the tribes) for these two measurements (Fig. 20), we find that, as expected, the Ami as one group and the Paiwan and Rukai as the other, are located at opposite extremes of the diagram, with the other tribes between them.

Table 15. Means and standard deviations of anthropometric measurements for men of the Taiwan aboriginal tribes (units: head measurements, mm; body measurements, cm).

Measure-ment	Atayal		Saisiat		Bunun		Tsou		Paiwan		Rukai		Puyuma		Ami	
	Mean	s.d.	Mean	s.d.	Mean	s.d.	Mean	s.d.	Mean	s.d.	Mean	s.d.	Mean	s.d.	Mean	s.d.
HL	179.8	5.87	183.1	6.12	180.9	9.31	182.9	6.40	177.9	6.69	178.5	6.98	184.0	6.24	187.6	6.74
HB	145.7	4.88	146.0	7.90	148.2	4.58	147.5	4.83	150.1	6.36	150.4	7.06	149.6	6.02	145.1	5.81
HH	121.1	10.10	127.7	11.96	122.8	10.95	129.7	9.36	124.1	9.89	123.2	10.63	124.4	8.15	123.9	9.60
FB	102.5	6.41	102.8	4.17	105.7	5.32	105.1	4.11	102.7	5.01	103.9	5.12	103.6	4.22	104.5	4.70
ZB	134.3	7.33	137.8	5.21	134.3	6.73	137.4	7.26	140.8	5.72	142.2	7.43	142.2	5.96	141.2	5.06
BB	103.2	5.16	104.9	6.84	105.1	5.30	106.7	5.58	105.0	5.66	104.4	5.86	106.6	6.78	107.4	6.07
OB	31.4	2.76	31.5	3.89	30.1	2.77	31.9	2.41	32.1	2.63	32.3	2.91	32.7	3.00	32.2	2.33
FL	125.9	7.70	129.9	7.40	123.5	5.52	126.3	5.97	122.3	6.76	121.0	5.87	123.3	6.80	126.5	6.47
NL	52.1	4.74	53.8	4.49	50.6	3.46	53.4	4.58	51.7	3.51	51.3	3.75	52.7	3.58	52.4	3.49
NB	40.9	4.20	39.7	2.44	40.7	3.86	39.3	3.14	40.9	3.07	41.4	2.36	41.5	3.09	41.9	2.63
MB	51.0	4.14	47.7	4.04	48.1	3.69	50.3	5.00	49.0	3.51	51.1	3.81	51.3	3.58	52.4	4.05
ST	160.1	5.59	158.7	5.35	157.2	5.56	164.0	5.44	156.6	5.56	158.2	5.81	160.0	5.11	164.6	5.61
SN	130.6	4.58	128.4	4.75	128.5	4.93	133.3	5.08	128.0	4.73	129.5	5.07	130.9	4.76	134.4	5.22
IH	90.6	4.27	89.6	4.12	88.9	4.42	92.6	4.35	88.6	3.94	89.7	4.35	89.6	3.78	93.9	4.57
SH	83.5	3.14	83.5	3.19	81.9	2.71	86.0	3.04	82.5	2.63	83.3	2.88	85.6	2.86	85.4	3.38
BD	36.7	2.14	37.4	1.73	36.9	1.88	37.2	1.92	36.2	1.73	35.8	1.95	36.6	1.98	36.7	1.89
CD	26.2	1.40	26.5	1.16	27.2	1.15	26.2	1.41	25.6	0.30	25.4	1.23	26.1	0.30	27.1	1.08
CG	89.4	4.42	88.1	2.59	91.7	2.38	89.2	2.57	86.6	2.34	86.4	2.65	86.4	2.70	83.9	2.94
UL	72.9	6.16	70.5	3.19	70.5	3.26	74.5	3.65	71.5	3.16	71.9	3.24	73.3	3.15	75.7	3.86

Table 16. Means and standard deviations of anthropometric measurements for women of the Taiwan aboriginal tribes (units: head measurements, mm; body measurements, cm).

Measure-ment	Atayal		Saisiat		Bunun		Tsou		Paiwan		Rukai		Puyuma		Ami	
	Mean	s.d.	Mean	s.d.	Mean	s.d.	Mean	s.d.	Mean	s.d.	Mean	s.d.	Mean	s.d.	Mean	s.d.
HL	174.2	5.55	178.0	6.25	173.4	6.72	173.1	5.92	172.4	6.12	172.8	7.01	174.5	6.38	178.6	5.78
HB	141.4	6.23	142.3	4.92	142.5	6.19	142.3	5.08	144.4	6.28	144.9	5.72	141.0	7.11	138.3	6.58
HH	119.1	10.86	121.5	11.27	116.9	9.20	120.6	10.72	121.5	9.83	122.1	11.45	120.0	11.09	118.6	8.65
FB	99.7	4.62	100.4	4.64	103.3	5.11	101.9	3.29	101.5	6.81	102.0	4.94	101.1	4.07	101.1	4.58
ZB	128.8	5.98	129.3	7.32	126.2	6.06	131.5	5.22	134.4	4.64	134.8	5.44	134.3	5.98	133.7	4.76
BB	98.5	4.63	97.5	12.61	100.1	4.62	100.1	4.94	100.3	7.62	100.2	5.10	99.8	4.88	100.9	7.33
OB	30.4	2.74	30.6	3.45	29.9	3.12	29.7	2.51	32.5	3.86	32.1	2.74	31.9	2.69	31.4	2.58
FL	119.4	7.19	121.8	7.39	116.3	5.67	120.2	5.93	115.8	6.10	113.9	6.53	115.7	6.16	117.3	6.70
NL	49.3	4.36	50.8	4.20	46.4	4.69	50.4	3.69	48.3	3.92	48.2	3.36	48.7	3.49	49.4	3.63
NB	37.8	3.32	36.8	3.29	38.4	3.71	36.5	2.61	38.4	2.81	38.1	2.74	38.5	2.68	38.9	2.22
MB	47.5	4.16	44.8	4.40	45.4	4.10	48.0	3.99	47.6	2.87	47.7	3.46	48.1	3.07	49.1	3.02
ST	149.8	4.74	150.4	5.75	146.2	5.08	153.9	4.70	148.0	4.57	148.7	4.63	149.6	5.32	155.9	5.88
SN	122.7	4.14	121.3	5.20	120.0	4.22	125.3	4.09	120.6	4.17	121.2	4.25	122.0	4.64	127.3	4.67
IH	86.2	3.99	86.1	3.81	82.9	3.66	87.3	3.51	82.8	4.32	84.1	3.65	84.7	4.24	89.3	4.01
SH	79.8	2.93	79.4	3.29	78.1	2.76	81.8	4.30	79.1	2.87	79.9	3.05	80.9	3.21	81.5	3.35
BD	33.7	1.82	34.1	1.91	33.7	1.60	34.5	1.58	33.3	1.68	33.3	1.92	33.3	1.60	34.1	1.62
CD	26.1	1.51	26.9	1.46	27.0	1.32	26.6	1.50	26.0	1.06	25.9	1.14	26.1	1.75	27.0	1.15
CG	82.6	3.44	83.6	3.33	84.3	3.21	82.3	2.95	81.0	2.73	80.5	3.26	79.8	2.37	80.0	2.28
UL	68.0	3.62	66.8	2.72	65.7	2.42	70.3	3.00	67.5	2.41	67.5	2.77	68.1	3.70	71.3	3.81

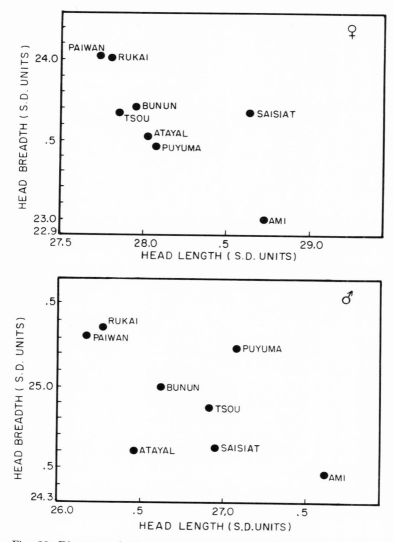

Fig. 20. Diagrams plotting the mean head breadth against the mean head length of each tribe in standard deviation units, showing the relationships between the tribes based on these two head measurements.

The Ami and Tsou rank high, and the Paiwan and Bunun low in stature, height of suprasternal notch, and spina ilica ventralis, a relative order which apparently goes along with the total upper extremity length and sitting height. The mean stature of the Ami, Tsou, and Puyuma falls in the lower range of the average category, and that of the remaining tribes in the short category.

A two-dimensional diagram was constructed for each sex (Fig. 21), using as one variable the means of the intercristal distance to represent a crossward body measurement, and stature to represent a longitudinal measurement, variation among the tribes in both these measurements being relatively large. Since these two measurements are correlated, the axes in the diagram are inclined at an angle $\cos^{-1} r$, where r is the correlation coefficient between the two variables; in this case the r value is 0.46 for men and 0.35 for women (Table 19). As on the head length and head breadth diagram, the standard deviation units, rather than the actual mean values, were used for the plotting. The distance between two points so plotted is equivalent, apart from a constant multiplier, to Mahalanobis' D^2 between two populations, which is discussed in the last section of this chapter (Rao, 1952). The tribes represented by these points are rather widely scattered in two-dimensional space, men more widely than women, as would be expected in view of the large between-group variance for these two measurements and the differences in magnitude between sexes as shown in the section on analysis of intertribal variance (Table 21). The Ami and Bunun show most separation between each other and from the remaining tribes; the Paiwan and Rukai, and the Puyuma and Atayal, appear relatively close.

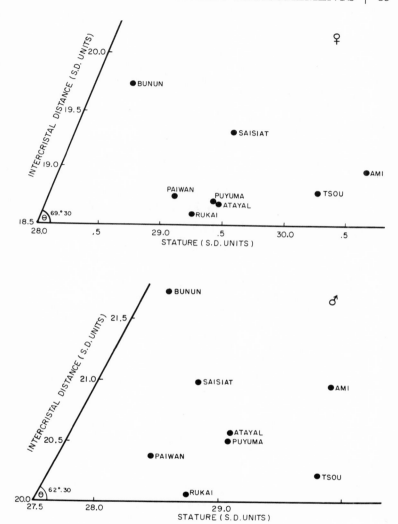

Fig. 21. Diagrams plotting the mean intercristal distance against the mean stature of each tribe in standard deviation units, showing the relationships between the tribes based on these two body measurements. The axes are inclined (see text for explanation.)

INDICES

Head and body indices may be considered measurements of the shape or form of the head and body. The following analyses show tribal and sexual variations.

Head Indices

Head indices were computed for each individual of each tribe according to the standard method:

$$\text{Cephalic index} = \frac{\text{Maximum head breadth} \times 100}{\text{Maximum head length}},$$

$$\text{Length-height index} = \frac{\text{Head height} \times 100}{\text{Maximum head length}},$$

$$\text{Breadth-height index} = \frac{\text{Head height} \times 100}{\text{Maximum head breadth}},$$

$$\text{Frontozygomatic index} = \frac{\text{Minimum frontal breadth} \times 100}{\text{Bizygomatic breadth}},$$

$$\text{Transversal cephalofacial index} = \frac{\text{Bizygomatic breadth} \times 100}{\text{Maximum head breadth}},$$

$$\text{Morphologic facial index} = \frac{\text{Morphologic facial length} \times 100}{\text{Bizygomatic breadth}},$$

$$\text{Jugomandibular index} = \frac{\text{Bigonial breadth} \times 100}{\text{Bizygomatic breadth}},$$

$$\text{Nasal index} = \frac{\text{Nasal breadth} \times 100}{\text{Nasal length}}.$$

The means and standard deviations are given in Tables 17 and 18.

Using the classification given in Comas (1960) we found that the mean cephalic indices of the eight tribes fall into

Table 17. Indices of head and body measurements for men of the Taiwan aboriginal tribes.

| | Tribe | | | | | | | | | | | | | | | |
| | Atayal | | Saisiat | | Bunun | | Tsou | | Paiwan | | Rukai | | Puyuma | | Ami | |
Index	Mean	s.d.	Mean	s.d.	Mean	s.d.	Mean	s.d.	Mean	s.d.	Mean	s.d.	Mean	s.d.	Mean	s.d.
Head																
Cephalic	0.81	0.033	0.80	0.050	0.82	0.054	0.81	0.042	0.84	0.051	0.84	0.044	0.81	0.046	0.77	0.039
Length-height	.67	.060	.70	.070	.68	.075	.71	.058	.70	.064	.69	.066	.68	.047	.66	.050
Breadth-height	.83	.065	.88	.088	.83	.076	.88	.066	.83	.073	.82	.064	.83	.064	.86	.072
Frontozygomatic	.76	.060	.75	.036	.79	.053	.77	.041	.73	.040	.73	.045	.73	.034	.74	.029
Transversal cephalofacial	.92	.046	.95	.060	.91	.045	.93	.058	.94	.040	.95	.054	.95	.040	.97	.033
Morphologic facial	.94	.057	.94	.055	.92	.062	.92	.057	.87	.055	.85	.065	.87	.063	.90	.053
Jugomandibular	.77	.051	.76	.045	.78	.045	.78	.048	.75	.044	.73	.049	.75	.053	.76	.041
Nasal	.79	.094	.74	.079	.81	.086	.74	.097	.79	.076	.81	.076	.79	.081	.80	.072
Body																
Rel. height spina ilica vent.	.57	.017	.56	.018	.57	.018	.56	.013	.57	.016	.57	.016	.56	.018	.57	.017
Rel. sit. height	.52	.016	.53	.015	.52	.017	.52	.016	.53	.016	.53	.013	.53	.012	.52	.019
Rel. intercristal distance	.16	.009	.17	.006	.17	.008	.16	.007	.16	.007	.16	.007	.16	.007	.16	.006
Rel. girth chest	.56	.043	.56	.025	.58	.079	.53	.023	.56	.023	.53	.025	.53	.033	.53	.028
Rel. length upper limb	.46	.034	.44	.015	.45	.018	.45	.015	.46	.017	.45	.015	.46	.017	.46	.019
Acromicristal	.71	.047	.71	.034	.74	.039	.70	.033	.71	.028	.71	.038	.71	.033	.74	.038

91

Table 18. Indices of head and body measurements for women of the Taiwan aboriginal tribes.

	Tribe															
	Atayal		Saisiat		Bunun		Tsou		Paiwan		Rukai		Puyuma		Ami	
Index	Mean	s.d.	Mean	s.d.	Mean	s.d.	Mean	s.d.	Mean	s.d.	Mean	s.d.	Mean	s.d.	Mean	s.d.
Head																
Cephalic	0.81	0.030	0.80	0.036	0.82	0.045	0.82	0.040	0.85	0.048	0.84	0.049	0.81	0.049	0.77	0.046
Length-height	.68	.063	.68	.070	.67	.058	.70	.065	.71	.067	.71	.072	.69	.068	.66	.047
Breadth-height	.84	.081	.85	.083	.82	.081	.85	.081	.84	.073	.84	.083	.85	.093	.86	.073
Frontozygomatic	.77	.041	.78	.050	.82	.063	.78	.033	.76	.050	.76	.033	.75	.040	.76	.030
Transversal cephalofacial	.91	.046	.91	.053	.89	.041	.92	.040	.93	.037	.93	.037	.95	.056	.97	.038
Morphologic facial	.93	.062	.94	.072	.92	.065	.91	.053	.87	.051	.85	.053	.86	.057	.88	.059
Jugomandibular	.77	.038	.75	.105	.79	.047	.76	.043	.75	.051	.74	.037	.74	.048	.75	.056
Nasal	.77	.094	.73	.080	.83	.107	.73	.069	.80	.086	.79	.078	.79	.071	.79	.072
Body																
Rel. height spina ilica vent.	.57	.018	.57	.016	.57	.016	.57	.013	.56	.021	.56	.016	.57	.020	.57	.017
Rel. sit. height	.53	.019	.53	.018	.53	.018	.53	.025	.53	.028	.54	.020	.54	.018	.52	.018
Rel. intercristal distance	.17	.011	.18	.009	.18	.009	.17	.009	.17	.008	.17	.007	.17	.011	.17	.006
Rel. girth chest	.56	.036	.56	.033	.58	.033	.53	.033	.56	.030	.53	.033	.53	.023	.51	.023
Rel. length upper limb	.45	.021	.44	.017	.45	.016	.46	.014	.46	.014	.45	.014	.45	.020	.46	.020
Acromicristal	.78	.048	.79	.052	.80	.044	.77	.047	.77	.041	.78	.044	.78	.048	.79	.045

two classes: mesocephalic (76 to 80.9)—Ami, Saisiat; brachycephalic (81 and over)—Paiwan, Rukai, Bunun, Atayal, Tsou, and Puyuma. The means for the Paiwan and Rukai are largest, and the mean for the Ami is smallest. There is almost no sex difference.

All the means of the length-height indices for the tribes fall into the hypsicephalic class (62.6 and over). The means for Rukai, Paiwan, and Tsou are highest; those for Ami, Puyuma, and Atayal, lowest. Sex differences in these indices vary in direction and magnitude among the tribes.

The means of breadth-height indices for the eight tribes fall into two classes: metriocephalic (79.0 to 84.9)—Rukai, Paiwan, Bunun, Atayal, Puyuma; acrocephalic (85.0 and over)—Ami, Saisiat, Tsou. The means for the Saisiat and Tsou tribes are the highest, whereas the mean for the Rukai is lowest. As in the length-height index, differences between sexes vary in direction and in magnitude among the tribes.

The mean transversal cephalofacial index is higher in males than in females in all the tribes except the Ami and Puyuma, where it is about the same for both sexes. Among the tribes the mean for the Ami is highest in both sexes, and that for the Bunun lowest.

The frontozygomatic index, relating to the prominence of the cheekbone, is highest for the Bunun tribe, and lowest in the Paiwan, Rukai, and Puyuma. In each tribe the measurement is from 1 to 3 percent greater for women than for men.

In relative order among the tribes the means for the jugomandibular index are highest for the Bunun tribe, with the Tsou and Saisiat next, and the Rukai lowest. Sex differences are slight and may be due merely to sampling variation.

The nasal index for all the tribes falls into the class of mesorrhine (70.0 to 84.9), but there is considerable intertribal variation. The means indicate that the Saisiat and Tsou tribes have the narrowest type of nose, the Bununs the broadest, and the rest are intermediate. In most tribes the mean for females is higher than that for males.

Body Indices

The following body indices were computed for each individual of each tribe according to the standard method:

$$\text{Relative height of spina ilica ventralis} = \frac{\text{Spina ilica height} \times 100}{\text{Stature}}$$

$$\text{Relative sitting height} = \frac{\text{Sitting height} \times 100}{\text{Stature}},$$

$$\text{Relative intercristal distance} = \frac{\text{Intercristal distance} \times 100}{\text{Stature}},$$

$$\text{Relative girth of chest} = \frac{\text{Chest girth} \times 100}{\text{Stature}},$$

$$\text{Relative length of upper limb} = \frac{\text{Upper limb length} \times 100}{\text{Stature}},$$

$$\text{Acromicristal index} = \frac{\text{Intercristal distance} \times 100}{\text{Biacromial diameter}}.$$

The means and standard deviations are given in Tables 17 and 18. The mean relative sitting heights and heights of spina ilica ventralis vary less between sexes and between tribes than the head indices. But the mean relative intercristal distance is greater for women than for men in each tribe.

The mean relative chest girth of the Bununs is highest among the tribes; that for the Ami, lowest. The Saisiat, Atayal, and Paiwan tribes rank just below the Bunun, and the Rukai and Puyuma slightly above the Ami. The relative chest girth is of adaptive significance, since the Bunun tribe inhabits the highest altitudes and the Amis live almost entirely near sea level. Except in the Ami tribe, where the means for women are lower than for men, there is no sex difference in the mean throughout the tribes.

The mean relative upper limb length characterizes both sexes of the Saisiat tribe as brachybrachion (short arm), and of the remaining tribes as metriobrachion (intermediate arm), differences between sexes being rather slight.

There are large sex differences in the acromicristal index; the mean for women is greater than that for men within each tribe, and there is also considerable intertribal variation. The mean for Bunun women is highest and for Tsou and Paiwan women lowest. The means for Bunun and Ami men are highest, whereas the mean for Tsou men is lowest.

To summarize, the linear measurements and indices reveal a great deal of contrast between the aboriginal tribes, especially if the Ami tribe is taken as one group and the Paiwan, Rukai, and Bunun as another. The Amis have large longitudinal measurements, such as head length, morphologic facial length, nasal length, and stature. The other three tribes are large in crosswise or bilateral measurements, such as head breadth, bizygomatic diameter, and intercristal distance. Thus the Amis are of the meso-cephalic type, comparatively tall and slender with relatively small chest girth, whereas the Paiwans, Bununs, and Rukais are brachycephalic, with prominent cheekbones, short but wide bodies, and relatively large chest girth. The Atayal, Saisiat, Tsou, and Puyuma tribes are similar to the Amis in some respects and to the Paiwans, Bununs, and Rukais in others. Sex differences appear more pronounced in body than in head measurements, as would be expected in view of the differences in physical activity and physiology between men and women. But there are so many measurements and indices it would not be practical to discuss each with respect to differences between tribes and between sexes. Over-all differences and similarities among the tribes are evaluated in the sections on Discriminant Analysis and Mahalanobis' Distance Analysis.

As we have discussed the rather dark skin color and short stature of the Bununs and Paiwans in the previous chapter, we should perhaps point out here that in view of the mean statures for these two tribes a small proportion of each theoretically falls within the range of the stature of Negritos. (The mean and standard deviations for the stature of the Bununs and Paiwans are about 157 ± 5.5 cm, and the upper limit for the stature of the Negritos is 150 cm (Coon, 1962).) Theoretically 7 percent of the Bununs and Paiwans with statures of 1½ standard deviations below their mean are within the upper limit of the stature of the Negritos.) The general body conformation of the Bununs and Paiwans is proportional, their heads are brachycephalic, their noses short, and their faces round. These features are in common with the Negritos but should not be considered characteristic, since they are also typically Mongoloid. Figure 22 provides a visual picture of the relative stature and skin color with reference to the Chinese.

The Bununs' large chest girth should also be noted; being short in stature, their average index in this respect is even greater than that for the other tribes. I would attribute the size of the chest girth of the Bununs to the effect of natural selection. This tribe lives at the highest average altitude, and it is well known (Coon, 1962) that people living at high elevations tend to develop large chests to compensate for reduced oxygen tension. Note that the other western tribes, such as Atayal, Saisiat, and Paiwan, many of whom also live at rather high altitudes, are next largest in chest girth. The Amis and Puyumas, at only slightly above sea level, have the smallest chest-girth index. Further discussion of the genetic basis for the above are given in the concluding chapter.

CORRELATION BETWEEN MEASUREMENTS

To determine whether one observed variable in an individual was associated with any other, correlation coefficients were computed for each sex, treating the eight tribes as a single population. This procedure confounded the correlations between tribes and those between individuals within the tribes, but it greatly simplified the computation and presentation. A correlation coefficient between mean measurements of each sex was similarly computed for each tribe. The correlation coefficients between individuals are given in Table 19 and between the means in Table 20. Those for men and women are generally consistent, but in some instances the correlations based on individual measurements are quite different from those based on the tribal means.

In the head measurements the correlation between mean head length and mean head breadth is negative and significant in women, and negative and rather high, although not significant, in men. This indicates that as head length increases, head breadth decreases, affecting the shape rather than the total dimensions. Correlations based chiefly on individual measurements exist between bilaterals such as head breadth, least frontal breadth, bizygometric diameters, and bigonial breadth, and also between longitudinal measurements, such as head length, morphologic facial length, and nose length. All the correlation coefficients between interocular internal breadth and bizygometric diameter are significant.

All body measurements are significantly correlated except chest girth with sitting height and upper limb length for the

Fig. 22 (*left*). The relative stature and skin color of a Bunun man (*center*) with reference to Chinese. On his right is the local school principal and on his left the author (5′9″), both of whom are northern Chinese. The school is in the background.

Fig. 22 (*right*). The relative stature of the Paiwan with reference to Chinese. In the middle are a husband and wife of the Paiwan tribe. On their left is Mr. T. M. Lin, a southern Chinese. Note the compact and wide body build of the man, a quite common body form for the Paiwans. Their skin color is not as dark as that of the Bunun man in Fig. 22 (*left*).

Table 19. Correlation coefficients between anthropometric measurements of individuals for all eight tribes; values above the main diagonal are for men, those below for women.

	Head											Body							
	HL	HB	HH	FB	ZB	BB	OB	FL	NL	NB	MB	ST	SN	IH	SH	BD	CD	CG	UL
Men																			
HL	-0.05	-0.02	.07	.20[a]	.17	.18	.10	.33[a]	.23[a]	.22[a]	.24[a]	.33[b]	.32[b]	.26[b]	.26[b]	.25[b]	.32[b]	.10	.23[b]
HB	.02	.19[a]	.19[a]	.29[b]	.39[b]	.17	.11	-.01	.02	.10	.08	.07	.08	.01	.12	.15	.04	.11	.03
HH	.09	.09	.45[b]	.14	.12	.02	.04	.08	.02	.04	.03	.14	.05	.04	.11	.14	.05	.05	.00
FB	.15	.27[b]	.14	—	.26[b]	.22[a]	.16	.10	.03	.11	.10	.18	.18	.13	.14	.18	.21[a]	.15	.12
ZB	.09	.37[b]	.12	.26[b]	—	.32[b]	.26[b]	.09	.13	.22[a]	.26[b]	.22[a]	.24[a]	.17	.28[b]	.18	.12	.04	.17
BB	.09	.11	.06	.22[a]	.28[b]	—	.12	.10	.09	.11	.17	.23[a]	.28[b]	.21	.19	.22[a]	.21	.13	.19[a]
OB	.10	.18	.04	.16	.08	.12	—	.10	.10	.19	.16	.11	.09	.04	.12	.06	.05	.10	.09
FL	.28[b]	.05	.05	-.01	.08	.01	.04	.45	.41[b]	.04	.10	.25[b]	.24[b]	.22[a]	.16	.21[a]	.11	.14	.17
NL	.18	.03	.04	.12	.16	.08	.19	.41[b]	—	.08	.17	.20[a]	.17	.16	.15	.13	.02	.00	.18
NB	.14	.10	.05	.12	.16	.08	.19	.45	.07	.31	.40[b]	.07	.08	.01	.06	.10	.45[a]	.06	.03
MB	.13	.04	.09	.02	.22	.06	.18	.11	.19	.31	—	.23[a]	.23[a]	.18[a]	.18	.06	.03	-.01	.19[a]
Women																			
ST	.24[a]	.04	.13	.11	.31[b]	.23[a]	.11	.25[b]	.20[a]	.07	.23[a]	—	.93[b]	.83[b]	.68[b]	.47[b]	.46[b]	.21[a]	.65[b]
SN	.24[a]	.04	.06	.11	.29[b]	.22[a]	.09	.24[b]	.17	.08	.23[a]	.91[b]	—	.84[b]	.64[b]	.44[b]	.46[b]	.21[a]	.67[b]
IH	.20[a]	-.03	.04	.06	.19[a]	.12	.04	.22[a]	.16	.01	.18[a]	.79[b]	.82[b]	—	.50[b]	.38[b]	.44[b]	.17	.63[b]
SH	.13	.04	.08	.12	.28[a]	.20[a]	.12	.16	.15	.06	.18	.52[b]	.50[b]	.40[b]	—	.41[b]	.28[b]	.16	.45[b]
BD	.13	.12	.07	.22[a]	.25[b]	.19[a]	.06	.21[a]	.13	.10	.06	.47[b]	.46[b]	.41[b]	.30[b]	—	.47[b]	.40[b]	.31[b]
CD	.17	.02	-.04	.14	.06	.10	.05	.11	.02	.45[a]	.03	.35[b]	.37[b]	.34[b]	.20[a]	.37[b]	—	.31[b]	.36[b]
CG	.07	.17	.02	.24[a]	.12	.20[a]	.10	.14	.00	.06	-.01	.24[a]	.25[b]	.24[a]	.19[a]	.46[b]	.35[b]	—	.11
UL	.14	.00	.05	.06	.25[a]	.13	.09	.17	.18	.03	.19[a]	.65[b]	.67[b]	.60[b]	.38[b]	.35[b]	.29[b]	.10	—

[a]Significant at 5 percent level. [b]Significant at 1 percent level.

Table 20. Correlation coefficients between mean anthropometric measurements of each of eight tribes; values above the main diagonal are for men, those below for women.

	Head											Body							
	HL	HB	HH	FB	ZB	BB	OB	FL	NL	NB	MB	ST	SN	IH	SH	BD	CD	CG	UL
HL		−0.60	.33	.32	.17	.79[a]	.20	.56	.51	.12	.37	.76[a]	.73[a]	.72[a]	.71[a]	.52	.66	−0.19	0.64
HB	−.73[a]		−.09	.02	.48	−.09	.28	−.85[b]	−.44	.17	−.10	−.58	−.47	−.63	−.25	−.67	−.67	−.25	−.36
HH	−.15	.57		.20	.09	.49	.18	.50	.76[a]	−.75[a]	−.27	.38	.24	.29	.16	.59	−.01	−.02	.14
FB	−.45	.33	−.28		−.16	.52	−.37	−.14	−.28	−.08	.03	.32	.37	.33	.16	.19	−.01	.42	.22
ZB	−.07	.08	.60	−.10		.46	.86[b]	−.38	.10	.51	.46	.09	.17	.08	.72	−.56	−.47	−.94[b]	.30
BB	−.30	.08	−.21	.61	.49		.38	.14	.09	−.48	.66	.65	.67	.59	.63	.24	.37	−.33	.64
OB	−.09	.24	.50	−.11	−.51	−.69		−.18	−.67	.58	.41	.33	.37	.26	.30	−.42	−.58	−.92[b]	.52
FL	.47	−.30	.01	−.50	−.51	−.48	−.67		.76[a]	−.52	−.76[a]	.44	.26	.42	.66	.87[b]	.50	.20	.13
NL	.51	−.28	.47	−.68	.14	−.48	−.18	.76[a]		−.52	.52	.52	.36	.40	−.02	.63	.00	−.23	.32
NB	.02	−.24	−.40	.19	.31	.58	.58	−.76[a]	−.52		.62	−.04	.14	.04	.04	−.69	−.01	−.52	.26
MB	−.07	−.35	.07	−.15	.72[a]	.66	.41	−.37	.10	.39		.64	.76[a]	.64	.64	−.41	−.12	−.54	.84[b]
ST	.58	−.68	.02	−.33	.30	.19	−.15	.41	.69	−.18	.57		.98[b]	.97[b]	.87[b]	.36	.34	−.16	.92[b]
SN	.49	−.74[a]	−.16	−.25	.26	.33	−.18	.29	.53	−.04	.67	.97[b]		.96[b]	.85[b]	.19	.30	−.22	.97[b]
IH	.70[a]	−.77[a]	−.10	−.46	.09	.02	−.30	.53	.71[a]	−.21	.43	.95[b]	.93[b]		.74[a]	.28	.37	−.17	.90[b]
SH	.27	−.52	.17	−.27	.52	.29	.00	.18	.61	−.16	.74[a]	.87[b]	.86[b]	.77[a]		.26	.05	−.39	.84[b]
BD	.45	−.45	−.17	−.08	−.36	−.16	−.74[a]	.79[a]	.63	−.63	−.07	.70[a]	.65	.73[a]	.46		.63	.52	.03
CD	.63	−.54	−.55	.23	−.55	−.06	−.61	.48	.12	−.14	−.37	.36	.34	.43	.00	.71[a]		.49	.15
CG	−.04	.19	−.33	.18	−.93[b]	−.54	−.78[a]	.55	−.06	−.51	−.84[b]	−.36	−.38	−.20	−.59	.38	.50		−.38
UL	.34	−.61	.01	−.24	.47	.43	.00	.16	.52	−.01	.80[a]	.94[b]	.97[b]	.84[b]	.91[a]	.52	.14	−.55	

[a]Significant at 5 percent level. [b]Significant at 1 percent level.

individuals, and the means of some bilateral with longitudinal measurements. The correlation coefficients between some parts of the body appear to be meaningless, since one measurement is contained in the other, e.g., stature contains suprasternal notch height, sitting height, and height of the spina ilica ventralis, so that all are highly dependent measurements. But these values were computed for comparison with other correlation coefficients.

Careful examination reveals some interesting linear relationships between parts of the human head and the body. Ordinarily we assume that people of tall stature will have large body parts, and this appears to be true for most parts of the body, though not for all parts of the head as based on individual measurements. Head length, bizygomatic diameter, bigonial breadth, and morphologic facial length are correlated with most body measurements, but head breadths and frontal breadth are not. Kroeber (1948), in discussing the head form and stature of the Negritos, mentioned that "their broadish short heads might well be a function of their short stature." This relationship apparently holds true for the Taiwan aboriginal tribes.

A correlation between two measurements of an individual comprises both the genetic and environmental, each of which includes both individual and tribal components. Assuming the effect of environment on an individual's physical measurements to be random either in a minus or a plus direction, the mean phenotype of a character is the best indicator of the genotypic mean of a population in a given environment. Correlation between the means of one character and another in the same population is therefore an estimate of both genetic and environmental correlations, but the environmental component is for the tribes only.

In view of the evolutionary skeletal changes in the human and other species, skeletal variations constitute the main criterion for classification of species and subspecies, and the correlations here observed contain genetic components known to exist in the human (Newman et al., 1937) and other (Bailey, 1956) species. Genetic correlation is an indication of pleiotropism or linkage of the genes concerned, or the presence of supergenes (Ford, 1964). Perhaps in the course of evolution such genetic properties safeguard the development of organisms into fully characterized biologic individuals.

Coefficients of correlation between the physical measurements of human beings have been reported by Majumdar and Rao, and other Indian workers (Majumdar and Rao, 1960) based on data from different Indian Muslim and non-Muslim groups. Howells (1951) also reported correlation coefficients from the measurements of 152 students at the University of Wisconsin, U. S. A. The correlations reported by these investigators are fewer than those we computed. Majumdar and Rao computed the pooled correlations for all the Muslim groups, the non-Muslim groups, and the combined groups. Comparing their values with ours, based on individual measurements, we find that theirs are generally higher within the head and body measurements, but those between them are generally comparable with ours. By and large the correlation coefficients of the American data are greater than those of the Indian. Environmental differences may account for the general difference in the correlations between the data from the three different sources.

INTERTRIBAL VARIANCE

For each anthropometric measurement the variance component ($\sigma_B{}^2$) for between-tribes was computed from the between-tribe mean squares and within-tribe mean squares. The fraction ratio of the between-tribe variance to the total variance was calculated for each measurement according to the formula σ_B^2/σ_T^2, where σ_T^2 is the total variance. These fractions are given in Table 21 for both men and women.

The general method for the computation of the between-tribe variance can be illustrated by the following example for variation of head length.

Source of variation	Degrees of freedom	Sum of squares	Mean squares	F
Between groups	7	8896.60	1270.94	26.23
Within groups	767	37156.40	48.44	

The between-group mean square is an estimate of $\sigma_W^2 + K_0\sigma_B^2$, where σ_W^2 is the within-group variance, σ_B^2 is the between-group variance, and K_0 is the harmonic mean of number of people in each tribe obtained according to the equation

$$K_0 = \frac{1}{n-1}\left[\Sigma K_i + \frac{\Sigma K_i^2}{(\Sigma K_i)^2}\right],$$

in which K_i is the total number of people in the ith tribe.

The within-group mean square is an estimate of the within-group variance only. Therefore

$\sigma_B^2 =$

$$\frac{\text{between-group mean square} - \text{within-group mean square}}{K_0}$$

$$= 12.77$$

and

$$\sigma_W^2 = \text{within-group mean square} = 48.44.$$

Since the total variance $\sigma_T^2 = \sigma_B^2 + \sigma_W^2$,

$$\frac{\sigma_B^2}{\sigma_T^2} = \frac{\sigma_B^2}{\sigma_B^2 + \sigma_W^2} = \frac{12.77}{12.77 + 48.44} = 0.209.$$

In the head measurements the variance fraction for length, breadth, bizygomatic diameter, morphologic facial

Table 21. Variance fraction[a] between the eight Taiwan aboriginal tribes for each anthropometric measurement. Each of the between-tribe variances as tested against the within-tribe variance is highly significant ($P < 0.01$).

Measurement	Men	Women	Measurement	Men	Women
HL	0.209	0.125	ST	0.248	0.257
HB	.111	.114	SN	.206	.219
HH	.047	.021	IH	.186	.222
FB	.042	.041	SH	.198	.112
ZB	.208	.251	BD	.056	.042
BB	.049	.016	CD	.256	.122
OB	.063	.100	CG	.093	.096
FL	.140	.110	UL	.203	.224
NL	.050	.084	Body weight	.189	.060
NB	.060	.050			
MB	.145	.103			

[a] Degrees of freedom: for men, 767, 7; for women, 900, 7.

length, and mouth breadth, which range from 0.103–0.251, are greater than those for the other head measurements, which range from 0.016 to 0.100. The F test of between-group mean squares against within-group mean squares for each is significant at the 5 percent probability level. The variance percentage for the bizygomatic diameter is the largest, and the variance fractions for morphologic facial length and mouth breadth are comparable in magnitude to those for head breadth. There is a large sex difference in the variance fractions for head length and bigonial breadth measurements: those for men are greater than those for women, but the variance fraction for the bigonial breadth is rather small for both.

In the body measurements the variance fractions range from 0.04 in shoulder breadth to 0.26 in stature for women, and 0.06 in shoulder breadth to 0.25 in intercristal distance for men. As the variance fraction for stature in both men and women is almost the largest among the measurements, other measurements closely correlated with stature would be expected to have large variance fractions. This was found to be the case in our data. The large variance fraction for male intercristal distances indicates that this measurement has high classificatory value for men. The fact that this value is smaller for women is ascribed to physiologic differences between the sexes and to the reproductive history of women. We omitted from examination pregnant women at any visible stage, but it is difficult to ascertain early stages of pregnancy, and the difference in reproductive history among women is possibly the major factor causing an increase in σ_W^2, and thus a decrease in the fraction of the total variance of σ_B^2. Apparently upper limb length is a useful measurement in classification as indicated by the rather high variance fractions in both sexes.

The classical twin studies by Newman et al. (1937) showed that in stature the correlation coefficient between identical twins is 0.93 and between fraternal twins 0.64. These estimates agree with the recent twin study by Osborne and DeGeorge (1959). In other mammalian species, such as the mouse, genetic variance or heritability of body size expressed by body weight is also fairly high (Chai, 1956; Falconer, 1961). Stature and correlated body measurements in human beings are evidently highly heritable.

There is, however, considerable discrepancy between the relative order of fractions of between-tribal variance and the ratio of dizygotic to monozygotic twin variance in the studies of Osborne and DeGeorge (1959). For instance, for between-tribal variance interzygomatic diameter and head length give the highest fractions, whereas in the Osborne and DeGeorge study of these two measurements the ratio of dizygotic variance to monozygotic variance is about the lowest. There are also discrepancies between the results of body measurements. Apparently the relative order of genetic and environmental variance in anthropometric measurements is quite different between populations and between individuals within populations or ethnic groups.

Harrison (1961), doubtful that data from studies of twins can be applied to physical anthropology and evolution, commented on the work of Osborne and DeGeorge (1959) as follows: "More troublesome still is the absolute independence of the environmental heterogeneity affecting families and populations. Even the study of monozygotic twins reared apart, by far the best way of estimating environmental effects, gives but minimum value to the possible contribution that a particular environmental difference is making. And the effects of environmental variation depend not only on the magnitude of this variation but also on the nature of its range." In general Harrison's view is correct. For instance, in the head length measurement, considered by physical anthropologists a good index for racial classification, the findings of Osborne and DeGeorge showed an inconsistency between men and women. In effect, if a heritability computation is based on the variance Osborne and DeGeorge obtained for identical and fraternal twins, an unbelievably high negative value is obtained for men. But in some other measurements, such as for stature, large genetic variances are in agreement with those of this study and of Newman et al. as mentioned above. Thus Osborne and DeGeorge concluded: "In drawing conclusions from the analysis of interpair differences, one must remember that the magnitude of these differences is of an entirely different order from that encountered in other types of data, such as those from studies of unrelated individuals or different racial groups . . . The measurement differences found in our study characterize this specific twin sample; these differences are proportional to the magnitude of the genetic and environmental components of variability in this twin population only."

In studying similarities or differences between human populations based on anthropometric measurements, one always faces the fact that environmental variation and its contribution are difficult to estimate. The between-group variance fraction is not entirely genetic. The variance within groups contains probably an even greater environmental component. For all practical purposes it may be said that characters having a greater ratio of between-tribe variance to total variance have greater classificatory or discriminatory value as far as the aboriginal tribes are concerned. The ratio of the between-tribe variance to the total variance of a measurement may be considered as a classification coefficient of that measurement.

DISCRIMINANT ANALYSIS

Assuming the effects of genetic determinants on linear measurements of characters to be additive, use of the discriminant function originally derived by Fisher (1936) in taxonomic classification is appropriate for analysis of the anthropometric measurements of the aboriginal tribes. This method of examining the similarities or differences between groups pools all numerical measurements in a compound index. It was shown to be a powerful technic for handling multiple measurements in Fisher's application to botanical problems and to sexual classification of ancient human skeletons. Fisher derived the method for discriminating between two groups.

For discriminating among a number of populations (Rao, 1952), the total population space is divided into mutually exclusive regions, one for each population, and an observed member is allotted to its population region. The boundaries of the regions cannot be determined by a single discriminant function without sacrificing some discriminatory power, and to achieve optimal properties, several functions have to be computed (Kendall, 1957). For the present analysis one set of coefficients and a constant for each tribe have been derived in each run, based on the following general equation:

$$f_i\ (Z,\ Z_2,\ \ldots,\ Z_n) = \sum_{p=1}^{n} Z_p\ c_{pi} - C_{n+l,i},$$

where i is the number of tribal groups, Z_p the pth anthropometric measurements, c_{pi} the coefficient for the pth measurements of the ith tribe group, and $C_{n+l,i}$ is a derived constant for the ith tribe. For the method of computing coefficients and the constants see BIMD 04 of UCLA Computing Center Program, UCLA, Los Angeles, California, and Kendall (1957) as programming reference.

The measurements of each individual in a group were multiplied by the corresponding coefficients of one set and a derived constant was added. Similar computation was made with each of the other sets of coefficients on the measurements of the same person, and the corresponding constants were added. The largest compound value occurs in the group to which the coefficient belongs, and theoretically this is the group to which the individual belongs. The greater the number of individuals belonging to their own group, the greater is the divergence between groups, and the homogeneity of individuals within each.

Depending on number of variables and number of groups, the calculation of the discriminant function can be too complex for a desk calculator and even for some types of electronic computers. Analysis of our data, comprising 19 head and body measurements, was made on an IBM 7090 computer. Due to program and computer limitations we could not handle all eight groups in a single run; two runs were therefore made for each sex, the first including five tribes, and the second the remaining three tribes along with two of the tribes included in the first run. By cross-checking the results of the two runs for each sex, and between sexes, a picture of the differences or similarities between the eight tribes was obtained.

The number of individuals in a tribe was classified into the five tribes of a run and were converted to percentages by dividing each by the total number of individuals in that tribe. Since tribal size differs, this conversion was convenient for examination of the divergence and homogeneity of one group relative to the others. Four tables in matrix form are presented to show the distribution of five aboriginal groups for each sex (Tables 22–25). On the main diagonal of each of these matrix tables is the percentage of individuals classified into its own group, and the values off the main diagonal represent the percentages of individuals classified into other groups according to the discriminant function analyses.

Table 22.   Classification matrix (percentages) based on the discriminant-function analysis of 19 physical measurements for men of Taiwan aboriginal tribes.[a]

Group	Atayal	Bunun	Paiwan	Rukai	Ami	Total number of individuals
Atayal	53.1	15.6	10.4	9.4	11.5	96
Bunun	7.3	80.2	7.3	3.1	2.1	96
Paiwan	10.2	2.4	60.6	23.6	3.1	127
Rukai	7.2	0.9	27.0	55.9	9.0	111
Ami	6.8	2.7	4.1	4.8	81.5	146

[a] The values in the 5 × 5 matrix table are the percentages of people of each tribe classified to their own and the other tribes based on the discriminant-function analysis. The values on the main diagonal represent percentages of right classification and those off the diagonal misclassifications. For example, the Atayal tribe has 53.1 percent classified to its own tribe, and 15.6, 10.4, 9.4, and 11.5 percent classified into the Bunun, Paiwan, Rukai, and Ami tribes, respectively. The percentages should add up to 100. (Similar explanations apply to Tables 23–26.)

Table 23. Classification matrix (percentages) based on the discriminant-function analysis of 19 physical measurements for women of Taiwan aboriginal tribes.

Group	Atayal	Bunun	Paiwan	Rukai	Ami	Total number of individuals
Atayal	57.1	17.7	7.5	9.5	8.2	147
Bunun	11.8	81.8	2.7	3.6	0.0	110
Paiwan	12.7	2.0	50.7	30.0	4.7	150
Rukai	10.6	3.2	24.5	50.0	11.7	94
Ami	10.6	0.0	3.2	10.6	75.5	94

Table 24. Classification matrix (percentages) based on the discriminant-function analysis of 19 physical measurements for men of Taiwan aboriginal tribes.

Group	Atayal	Saisiat	Tsou	Puyuma	Ami	Total number of individuals
Atayal	58.3	11.5	15.6	11.5	3.1	96
Saisiat	9.1	68.2	10.1	7.6	4.5	66
Tsou	12.7	15.5	53.5	7.0	11.3	71
Puyuma	6.3	6.3	7.9	66.7	12.7	63
Ami	4.8	2.7	6.2	13.7	72.6	146

Table 25. Classification matrix (percentages) based on the discriminant-function analysis of 19 physical measurements for women of Taiwan aboriginal tribes.

Group	Atayal	Saisiat	Tsou	Puyuma	Ami	Total number of individuals
Atayal	49.7	13.6	17.0	12.2	7.5	147
Saisiat	11.7	70.0	3.3	10.0	5.0	60
Tsou	12.0	8.0	69.3	6.7	4.0	75
Puyuma	6.4	5.5	11.8	64.5	11.8	110
Ami	5.3	3.2	4.3	17.0	70.2	94

Evaluating relative gene flow between tribes may be considered a unique feature of the discriminant function analyses. (The test of significance can be made by chi-square statistics but it requires larger sample sizes than we have.) As the sample sizes are small, interpretation of our findings with respect to this point should be cautious. The Atayal group appears to be quite heterogeneous, as a comparatively large percentage of its population fell into other groups. Among the three tribes of Atayal, Saisiat, and Tsou relatively large numbers in each tribe were classified into the other two. In addition, a large percentage of Atayals were classified into the Bunun group, and vice versa. For any two aboriginal groups, if a larger percentage of people is classified into the other than vice versa, more migration or interflow of genes would be indicated in that direction. The close relationship between the Rukai and Paiwan tribes is shown by the large percentage (23 to 30 percent) of people of one group classified into the other. Among the eight tribes, 70 to 81 percent of the Bunun and Ami people are classified into their own groups. Largest of the tribes, they may also be considered the most homogeneous and isolated.

To check the efficiency of the discriminant function analysis on the one hand, and the relationship of some of the aboriginal tribes to ethnic groups in southeast Asia on the other, a computer run was made with the Taiwan Chinese, Jahai (aborigines indigenous to Malaya, measurements from Sebesta and Lebzelter, 1928) and the Puyuma, Paiwan, and Atayal tribes, based on 11 head measurements. Those in the Chinese group were descended from Fukianese and Cantonese immigrants from the mainland of China (Fig. 1). The results of this analysis are given in Table 26.

Although some individual groups were small and the data came from different sources, and though there may

Table 26. Classification matrix (percentages) based on discriminant analysis of 11 physical measurements for women of different ethnical groups.

Group	Chinese	Jahai	Puyuma	Paiwan	Atayal	Total number of individuals
Chinese	65.2	0	8.7	17.4	8.7	23
Jahai	0.0	91.5	3.4	3.4	1.7	59
Puyuma	14.5	3.6	37.3	30.0	14.5	110
Paiwan	14.0	2.7	22.0	48.7	12.7	150
Atayal	15.1	1.4	8.9	6.8	67.8	146

have been some difference in the technics of taking measurements, the distribution of the percentages (Table 26) in each of the five groups reveals interesting relationships. The fact that no Chinese were classified into the Jahai group, nor any Jahais into the Chinese, indicates a great divergence. The percentages of Puyumas, Paiwan, and Atayals classified into the Chinese group are much greater than those for the Jahai.

MAHALANOBIS' DISTANCE ANALYSIS

The application of squared generalized distance to group constellation problems in anthropology, taxonomy, and genetics, and even in biologic assay, has been reported by various investigators. The theory and its applications have been thoroughly discussed by Mahalanobis (1925) and Mahalanobis et al. (1949), and Rao (1955) in a series of papers. Rao (Majumdar and Rao, 1960), stating that it is the logical basis for discriminating between species, subspecies, and any subject groups, applied this method to interrelationships of Bengal groups in India. Recently Hanna (1962) used it to study relationships among Southwest Indian groups. For studying differences or similarities between groups it is a most useful statistical tool, since multiple variant measurements correlated to different extents can be handled. Perhaps because of the heavy calculations involved it has not been widely used.

The Mahalanobis distance is rather similar to, and is actually a type of, the discriminant function analysis used in the previous section. Both are based on derivation of a series of coefficients for the variables, so that the measurements of the different characters can be linearly combined. The generalized distance between two groups gives a single value that represents the biologic distance between them, whereas the discriminant analysis gives the number of subjects of one group classified as members of another. The more the subjects of a group are misclassified, the less is the generalized distance. The discriminant analysis shows the relative heterogeneity of different groups, i.e., when a greater percentage of subjects of a group is classified to a second group than is classified from the second to the first, it is implied that the first group is more heterogeneous than the second. Distance analysis would not show this. On the other hand, when several groups are studied, the distance between one and another provides a unique and clear picture of the relative relationship that would otherwise be difficult to visualize, especially for a large number of groups. Thus the two types of analyses supplement each other in providing a deeper understanding of genetic structure and relationship between groups in population studies.

The Mahalanobis D^2 between two populations is defined (Majumdar and Rao, 1960) by

$$\Delta^2 = \Sigma\Sigma\lambda^{ij} \, \delta_i\delta_j,$$

where λ^{ij} is the matrix reciprocal to the dispersion matrix and δ is the difference between the mean values of ith or jth character of the two populations. The estimate for Δ^2 is the D^2 statistic

$$D^2 = \Sigma\Sigma s^{ij} \, d_id_j,$$

where s^{ij} is the sample estimate of λ^{ij} and d of δ. The dispersion matrix is obtained by pooling estimates from within groups. It is a 19th-order matrix in the present case.

For testing the significance of the D^2, the F statistics are computed according to the following equation

$$F(n, n_1 + n_2 - 1 - n) = \frac{n_1 n_2(n_1 + n_2 - n - 1)}{n(n_1 + n_2)(n_1 + n_2 - 2)} \cdot D^2,$$

where n is the number of the different anthropometric measurements and n_1 and n_2 are the sample sizes of two tribes.

The procedure for machine calculation is outlined in BIMD 05 of the University of California Los Angeles Computing Center (Rao, 1952), and the results are given in tables with triangle matrices, one for men (Table 27) and one for women (Table 28). Figures on the last line of each table represent average distances from each tribe to all the others, obtained by adding all seven D^2 for a tribe and dividing by seven.

The squared distances between the Atayal and the remaining tribes range between 2.3 and 5.6 for women and 3.0 and 6.2 for men. Among them, the smallest distance is

Table 27.   Biologic distance (Mahalanobis D^2) between the men of the Taiwan aboriginal tribes.

	Atayal	Saisiat	Bunun	Tsou	Paiwan	Rukai	Puyuma	Ami
Atayal		3.9	3.9	3.0	3.9	3.8	4.3	6.2
Saisiat			5.7	4.3	5.5	9.0	6.7	8.6
Bunun				9.3	7.1	9.1	9.6	10.0
Tsou					6.4	6.6	5.3	6.1
Paiwan						1.1	3.2	9.0
Rukai							2.5	7.1
Puyuma								4.6
Average	4.2	6.3	7.8	5.9	5.2	5.6	5.2	7.4

Table 28. Biologic distance (Mahalanobis D^2) between the women of the Taiwan aboriginal tribes.

	Atayal	Saisiat	Bunun	Tsou	Paiwan	Rukai	Puyuma	Ami
Atayal		3.2	3.9	2.3	3.6	4.1	3.3	5.6
Saisiat			6.3	7.0	8.6	9.4	7.7	9.4
Bunun				8.1	8.1	8.8	8.6	16.0
Tsou					5.0	4.8	3.7	5.3
Paiwan						0.5[a]	1.3	6.9
Rukai							0.8[a]	5.4
Puyuma								3.6
Average	3.7	7.7	8.5	5.2	4.9	4.8	3.6	7.5

[a] Not significant.

between the Atayal and Saisiat groups in women and the Tsou in men, and the largest distance for both sexes is between the Atayal and Ami tribes The squared distances between the Saisiat and other groups are larger than for the Atayal. All the squared distances between the Bununs and the other tribes are relatively large; and that between Bunun and Ami is largest for both men and women. In all except the Atayals, distances are above 6.3 for women and 5.7 for men. Distances between the Tsou and the other tribes are rather variable, the largest being between it and Bunun, 8.1 for women and 9.3 for men. It is interesting to see that the distance between Paiwan and Rukai is not significant in women and is very small in men. The distances of these two tribes from the others are variable in the same ratio. The distance between the Puyuma and Rukai tribes is not significant, and very small for women between the Puyuma and Paiwan. For the Ami group the distances are in general rather large.

The average distance of each tribe from the others may be considered as the measurement of relative isolation or genetic homogeneity. Since a large distance between two groups means that few people in one tribe resemble those of the other, it also means that there are less genes in common between the two groups or that the rate of interflow of genes is small. The Ami and Bunun tribes have the highest average distances from all others and the Atayal tribe has the lowest.

A point diagram (Fig. 23) has been prepared showing the "biologic distances" between each pair of the eight aboriginal tribes and based on the value of D^2 obtained for both sexes. This diagram should be considered only an approximation of the true biologic differences or similarities. If the notion of hyperspace associated with the Mahalanobis D^2 is kept in mind, the difficulty inherent in an attempt to express its values in two dimensions can be appreciated. The differences in tribal position and geographic distribution are clearly shown in this diagram. The Bunun and Atayal tribes have exchanged their relative geographic positions. The Saisiat and Bunun are located in the far left corner and the Ami is in the upper right. The

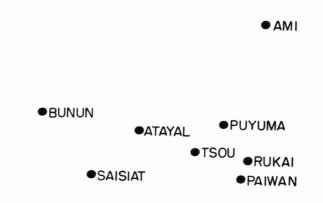

BIOLOGICAL DISTANCE BETWEEN TRIBES

Fig. 23. A scatter diagram showing approximate biologic distances between the tribes (see text for explanation).

remaining tribes are clustered together, with the Rukai and Paiwan staying the closest.

To summarize, the general pattern of biologic relationships among the eight Taiwan aboriginal tribes has some correlation with geographic distribution, particularly in respect to the extent of physical isolation. The Ami and Bunun tribes are most distinct from the others. The Paiwan and Rukai are more closely related with each other than with any other tribe. These findings agree with those obtained from the anthroposcopic and anthropometric data, with the author's subjective observation, and with the opinions of anthropologists (Kano, 1955). The Puyuma, an eastern tribe, appears to have a closer association with the Rukai, a western tribe, than with its eastern partner, the Ami, in the D^2 analysis. It is somewhat surprising that this close relationship between the Puyuma and Rukai as revealed by the anthropometric data seems not to have been previously claimed by anthropologists. The Atayal and Saisiat tribes are closely associated geographically and also in biologic distance, but their association is not so close as that of the Paiwan and Rukai. It is my subjective observation that the facial appearance, short stature, and compact body build of the Bununs give the impression that their most closely related tribe would be the Paiwan; but the D^2 analysis shows them to be more closely related with the Atayals and Saisiats. Leaving the question of possible origins for discussion in the last chapter, suffice it to say here that geographic isolation, with social isolation superimposed upon it, has a strong bearing on biologic relationships. Both geographic and social isolation appear basic to tribal variations. The characters here observed and measured are unquestionably determined by multiple factors or polygenes, and Mahalanobis' distances here obtained between the tribes therefore represent their genetic resemblances.

♀♂

CHAPTER 7.   TASTE THRESHOLD FOR PHENYLTHIOUREA

A test for taste sensitivity to phenylthiourea (phenylthio-carbamide, PTC) was given to adults in eight aboriginal tribes and to school children in five. Because we were pressed for time toward the end of our survey we were unable to give the taste test to school children in the Puyuma and Rukai tribes, and to only a small number in the Ami.

The test solution was made according to Harris and Kalmus (1949). Starting with a No. 1 concentration of 1.3 gm/lit, 13 others were prepared by diluting 50 percent each time, so that No. 2 contained 0.65 gm/lit, No. 3, 0.33 gm/lit, and so on, and the series was numbered 1 to 14. Since the solubility of phenylthiourea is rather slight, No. 1 was about at saturation point. For increasing the concentration, therefore, we used undissolved crystals and called it concentration No. < 1. The tap water of the City of Taipei used in making the solutions was boiled, let stand until cool, and the precipitated impurities were removed by not using the bottom portion of the water.

The solutions were kept in 250 to 500 ml beakers, from which small amounts of each were poured into individual wine glasses for tasting. The subject started with the lower concentrations of 14 or 13 and went on through the higher until bitterness was tasted; he then was asked to try the next lower concentration again. The subject was often

asked to taste the next lower or higher concentrations re-
peatedly until there was no doubt as to his true threshold.
In these trials a sample of plain water was always included
as a check, and the subject rinsed his mouth after tasting
the bitterness, before retasting a lower concentration.

Preliminary to this test a trial of reliability was made on
the Saisiat pupils by testing the same child at the same
hour on two subsequent days. The results (Fig. 24) show

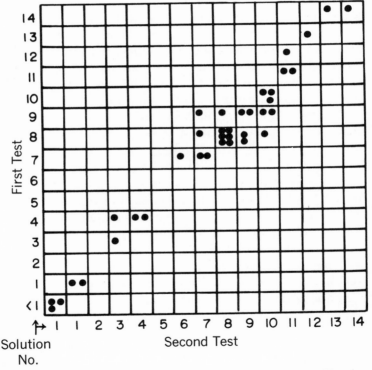

Fig. 24. The results of testing reliability of the PTC taste tests. Two tests
were given to the same individuals at the same time on two consecutive
days.

Table 29.   Mean PTC taste thresholds of school pupils and adults of different Taiwan tribes.

Tribes	School pupils				Adults			
	Male		Female		Male		Female	
	No.	Mean	No.	Mean	No.	Mean	No.	Mean
Atayal	21	6.28	21	7.76	106	7.32	171	7.92
Saisiat	47	8.26	48	7.69	67	7.24	61	7.63
Bunun	63	10.24	51	10.06	119	7.73	131	9.07
Tsou	30	10.03	31	10.06	82	10.18	78	10.23
Paiwan	57	9.00	61	9.10	142	9.57	239	9.64

that of 35 school children tested, 22 (almost two-thirds) had the same threshold on both days; 11 had a threshold difference of one dilution step; and 2 had differences of two steps. This level of reliability was considered good.

Table 29 gives the mean threshold scores for each sex and for the combined children and adults in each tribe. There appears to be a threshold difference between sexes which, although slight, is consistent for pupils and adults throughout the tribes; but threshold differences between school children and adults are not consistent. According to Harris and Kalmus there is about one dilution step decrease in sensitivity for each 30 years increase in age. Our results, however, did not show a definitely lower threshold for children, perhaps because the children were at an age when their taste buds were not fully developed. A scattering diagram showing threshold against age at 5-year intervals for Atayal adults (Fig. 25) reveals no trend toward increase of threshold with advancing age. There might, of course, be some increase at very old ages, but it would not appear in our data because there were extremely few people over 60.

Since age effect on PTC taste threshold, if any, was

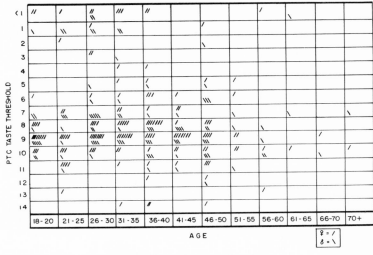

Fig. 25. PTC taste threshold with increase in age.

negligible, we pooled the data for school children and adults, but sexes were kept separate for comparison between the tribes. A small number of adult Chinese living in some of the tribal areas were included, as were Chinese children attending the same elementary schools as the aborigines. Most of these Chinese were Taiwan natives rather than recent immigrants from the mainland of China, and the results of their tests are given for comparison. Bimodal percentage distributions of the PTC threshold for each tribe are shown in Fig. 26. But in those tribes where there was a very small percentage with high thresholds, the mode of the left part of the distributions in the figure is not obvious. A comparatively large percentage with high thresholds occurs in the Atayal and Saisiat tribes, and the next largest in the Bunun; in the remaining five tribes the high-threshold percentage is very small. The percentage of high thresholds among the Chinese is smaller than that in the

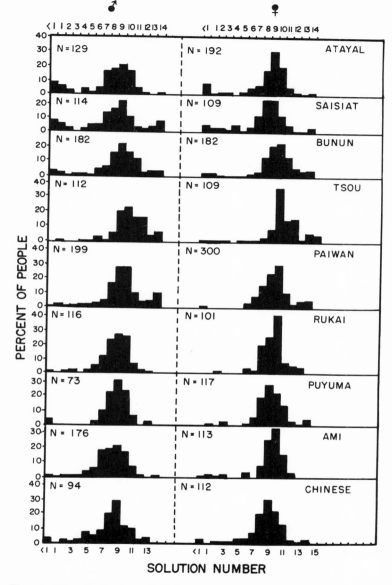

Fig. 26. The distribution of PTC thresholds in each of the eight aboriginal tribes.

Atayal and Saisiat tribes, comparable to the Bunun, and greater than in the remaining tribes.

We consider the anti-mode to be between thresholds 3 and 4, where it appears in the present data. People who tasted bitterness at concentrations 3 and above are considered nontasters. The distribution of tasters and nontasters is given in Table 30. The gene frequency is computed for each tribe, assuming that nontasters are homozygous recessive for a single gene and that breeding has been random in each tribe. The computation is made separately for men and women, since it is difficult to correct for sex differences. The average frequency for both sexes is considered a logical estimate. As in the distribution of nontasters among the different groups, the gene frequency for nontasters is highest in the Atayals and Saisiats, next highest in the Bununs,

Table 30. Distribution of nontasters and gene frequencies (q) in the aboriginal populations of Taiwan.

| Aboriginal groups | Percent nontasters[a] | | | | q[b] | | |
| | Men | | Women | | | | |
	No.	Percent	No.	Percent	Men	Women	Average
Atayal	127	16.8	192	12.0	0.40	0.35	0.37
Saisiat	114	15.0	109	11.9	.39	.35	.37
Bunun	172	8.8	182	5.5	.30	.23	.27
Tsou	112	0.9	109	3.7	.09	.19	.14
Paiwan	199	3.5	300	0.9	.12	.09	.11
Rukai	116	3.5	101	1.0	.12	.10	.11
Puyuma	73	4.1	117	3.5	.20	.12	.16
Ami	176	4.5	113	5.2	.21	.23	.22
Average	1089	7.1	1223	5.5	.26	.24	.25
Chinese	93	6.4	112	6.5	.25	.26	.25

[a] People who tasted bitterness at concentrations 3 and above are considered nontasters.

[b] Nontasters are assumed to be homozygous recessive for a major gene.

Amis, and Chinese, and lowest in the remaining tribes. If there had been inbreeding, especially in the small tribes, the gene frequencies for nontasters as shown in the table would be overestimated.

There is an interesting difference in the mode in the right part of the distribution, where the majority of the subjects belong. The Atayal and Saisiat tribes have a tendency to shift to the left; the Bunun, Tsou, and Paiwan, to the right. For comparing distribution with respect to threshold, the mean for the right-hand portion of the distribution with the cut-off point between concentrations 3 and 4 was computed for each tribe (Table 31). The mean for the Tsou tribe is highest; for the Paiwan, Bunun, Puyuma, and Rukai next; for the Atayal, Saisiat, and Ami tribes, and the Chinese group, lowest. The difference between the Tsou mean and that of any other group is significant, as are most of the differences between the next highest and the lowest groups.

These results raise many interesting questions. The first concerns the genetic basis of the PTC taste-threshold difference. Although the work of Harris and Kalmus (1952) did not fit the single-gene hypothesis, the clear-cut bimodal distribution for the results of many investigators as reviewed

Table 31.   Mean PTC taste threshold for the tasters of each tribe.

Tribe	No. of people	Mean ± standard error
Atayal	276	8.67 ± 0.120
Saisiat	193	8.70 ± .154
Bunun	338	9.59 ± .106
Tsou	216	10.36 ± .122
Paiwan	489	9.65 ± .080
Rukai	212	9.15 ± .102
Puyuma	183	9.29 ± .111
Ami	275	8.84 ± .098
Chinese	195	8.78 ± .136

by Saldanha and Nacrur (1963) indicates that (1) the PTC taste threshold is a genetic character and (2) there is certainly no hypothesis based on our present knowledge of genetics which would fit better than that of the single gene with modifiers. Comparing the anti-mode of distributions for the present data with some previously published results, the anti-modes of our results appear to be more to the left of the threshold distribution than those cited by Harris and Kalmus (1949). What causes this shift in the anti-mode is a question that may never be answered satisfactorily, but as likely an explanation as any is that the frequency of some genetic units affecting the physiology of taste varies among different populations. This explanation fits the assumption that differences in PTC taste thresholds are determined by minor genetic units as well as by a major gene.

The sex difference in PTC taste thresholds indicates that this trait is affected by the genetic environment of the individual, and supports the hypothesis that modifying genes exist that cause variation within the taster and the nontaster groups.

Whatever the genetic basis for the PTC threshold may be, the distribution of nontasters among the tribal groups appears to be correlated with geographic distribution. As shown in the map (Fig. 3) the Atayal tribe lives farthest north on the island, then in succession come the Saisiat, the Bunun, part of the Ami, and the other tribes toward the south. The percentage of nontasters is highest in the Atayal tribe, and diminishes in the Saisiat, Bunun, and other remaining tribes.

PTC taste is a dimorphic character the genetic mechanism for maintaining the balance of which is not clear. It has been claimed that some relationship exists between PTC taste and the development of goiter, and that nontasters are more likely to develop adenomatous goiter than tasters (Mourant, 1959). This claim is based on the fact that phenylthiourea and its related compounds contain the chemical group N—C=S and interfere with the normal synthesis of thyroxine. Some of these compounds occur in natural foodstuffs. But the basic factor or factors causing the variation in the ratio of nontasters and tasters among races and ethnic groups remain practically unknown. Some theories concerning possible evolutionary mechanisms involved in the percentage variations of nontasters among the aboriginal tribes may be outlined as follows.

Environmental differences correlated with geographic factors such as climate, soil, and natural foodstuff. Taiwan is a small island and theoretically should have little geographic variation, but actually the opposite is true, especially in winter when it is chilly and damp in the north, but relatively dry, sunny, and warm in the south. We have no definite knowledge concerning differences between northern and southern wild vegetation, upon which the aborigines still partly rely for food, but with so much contrast in climate it is likely that the type and availability of natural vegetation also varies. Since the aborigines inhabit *isolated* areas, natural conditions in one may be quite different from those in another. Selection pressure acting on tasters and nontasters may be a function of the natural environment, so that the gene frequency for the PTC taste may be different among tribes inhabiting different geographic niches.

Differences between gene pools in the tribes. Selection acting on the different genotypes of the PTC taste may be a function of a part or of the total genotype of an individual. The average genotype of the individuals in a tribe or the tribal gene pool may be a factor in determining a definite equilibrium of the three genotypes. Here it should be mentioned that the gene frequency of nontasters in the Chinese sample we obtained is similar to that obtained by Barnicot

(1951) from the Chinese in New York City, suggesting the possible importance of the gene pool in maintaining a given equilibrium of the PTC genes. Other genetic mechanisms that may be involved are the pleiotropic properties, if any, of the taster and the nontaster genes, and the adaptive properties of genes with which the taste genes are linked. Variation in selective value of characters that are correlated with PTC taste because of the mechanism mentioned above would influence the equilibrium of the PTC taste genes.

Genetic drift. If the tribes started from a heterogeneous population in ancient times and if the groups that branched off were small, the distribution of genes relating to PTC taste among the tribes might be subject to genetic drift. Ford (1964) pointed out that in ecologic populations selection ordinarily plays a far more important role than drift. However, since we are not at all clear about the operation of the selection mechanism and the magnitude of selection pressure, in tribal populations with respect to PTC taste, genetic drift as a possible cause of variations should not be rejected.

Migration. The aboriginal tribes are intrabreeding populations only in a relative sense. To a minor extent migration, causing interflow of genes between tribes, must have taken place.

It may well be that each of the evolutionary mechanisms mentioned above plays a role and that the variation in gene frequencies for PTC taste among the tribes is a consequence of their joint effect. The variation in gene frequencies as a polymorphic trait is further discussed in the concluding chapter.

As far as genetic relationship between the tribes is concerned, the results of our tests show that the Atayal, Saisiat, and Bunun tribes are closely associated. The extent of difference between the other tribes or between the western and eastern groups is not clear, but complete agreement between results relating to a single-gene trait, such as PTC taste, and those relating to multiple factors or polygenic traits, such as physical characteristics, could hardly be expected.

CHAPTER 8. BLOOD PRESSURE

Hypertension (high blood pressure or hyperplasia) was recognized as hereditary in the 16th century (Gates, 1946). Gates cited many cases and statistics showing family tendencies toward cardiovascular diseases, some of which were claimed to be caused by a single dominant gene, others by either a recessive or a dominant gene of irregular penetrance. Schwab et al. (1935) interpreted the higher incidence of cardiovascular diseases in Negroes over that in Caucasians as due to a hypersensitive vasomotor mechanism more frequent in Negroes than in whites.

Recent studies on the genetic basis of essential hypertension remain controversial. McKusick (1960) claimed that the polygenic theory makes better physiologic sense than the single-gene theory, since multiple factors are known to influence the level of blood pressure. He pointed out that correlation between the blood pressure of probands and first-degree relatives remains the same at all levels of pressure in the proband, a fact which fits into the polygenic hypothesis, as does the racial difference in blood pressure between Negroes and whites in the southern United States. Doyle and Fraser (1961), investigating the responses of hypertensive patients to norepinephrine, found that young adult normotensive sons of hypertensive parents have a significantly greater vascular response. They concluded that heightened vascular reactivity may depend on a specific defect inherited as a dominant characteristic. After

130

reviewing much reported evidence Schweitzer et al. (1962) concluded that it is not known to what extent genetic factors are responsible for essential hypertension. Platt (1963) having studied the siblings of patients with severe hypertension, believed that hereditary mechanisms are not explicable by multifactorial inheritance, but rather by the hypothesis of a single gene with incomplete dominance. In a study of 612 families randomly selected from two populations in South Wales, Miall and Oldham (1963) found evidence suggesting that arterial pressure is a continuously variable character, and they believe their data fit the hypothesis of multifactorial inheritance. They stated, however, that possibly single-gene inheritance plays a role in determining the arterial pressure of a very small fraction of this community.

Even less clear is the genetic basis for hypotension, a condition less frequently reported than hypertension. Garvin (1927) reported a family of five members with low blood pressure, and Kean (1944) observed hypotension in San Blas Indians. The average blood pressure of 408 of them was 105.2 mm Hg for systolic and 69.3 for diastolic, and a large portion of people showed marked hypotension. Murrill (1949), in a study of the natives of Ponape in the Eastern Carolines, found that the mean systolic and diastolic blood pressures for males, with ages ranging from 20 to 59 years, were averaging 111.9 and 76.4 mm, respectively; and for females, with ages ranging from 18 to 59 years, 111.0 and 76.6 mm, respectively. About 80 percent of males and females had systolic pressures below 120 mm. He cites cases of hypotension reported 30 years ago showing family or group patterns. In a survey of all the inhabitants of two Pacific Island communities, Gau in the Fiji and Abaiang in the Gilbert Islands, Maddocks (1961) reported that he found no substantially higher blood pressure in later life such as is characteristic of Western people, and that essential hypertension does not exist, or is rare, in these communities. Studying the influence of environment on essential hypertension, Cruz-Coke et al. (1964) compared blood pressure regression on age between the natives of Easter Island, an isolated Polynesian population, and a group of them which had migrated to continental Chile. They found the coefficient of regression on age 0.09 for the natives staying on the island and 0.34 for the emigrants. They concluded that civilization influences the relation of age to blood pressure.

Blood pressure varies among civilized populations. The average systolic pressure of young American male adults at age 20 is usually accepted as 120 mm Hg, the diastolic as 80, and the pulse pressure as the difference between the two, i.e., 40, as pointed out by Best and Taylor (1950). There is also a steady increase of blood pressure from adolescence to old age, and in men a rise averaging 0.5 mm in systolic and 0.2 in diastolic for each year of age after 20. Master et al. (1952) studied blood pressure of people employed by industrial plants in various sections of the United States and found a mean systolic pressure of 122.9 and a mean diastolic of 76.0 for men 20 to 24 years of age, and of 115.7 and 71.7 mm, respectively, for women of corresponding age. At ages 35 to 39, mean pressures were 127.1 and 80.4 for men, and 123.9 and 78.0 mm for women. In a comprehensive study of blood pressure in Bergen, Norway, Bøe et al. (1956) divided the population into two groups, and found some differences between them. At ages 20 to 24, average systolic and diastolic pressures for the two groups were 131 and 77 mm for men and 122 and 74 mm for women. In the age group 35–39, the average systolic and diastolic pressures were 134 and 81 mm for men and 130 and 79 for women. They found that females had lower blood pressure than males at young ages but greater rate of increase with advance of age. Thus at the age of 45 the mean blood pressures of females surpassed those of males. In view of the wide variation in blood pressure level and its regression on age between different populations, it is a question whether a universal "normal" can be adopted, or whether individual standards should be used.

As mentioned in Chapter 2, the blood pressure examination was the next-to-last item in the sequence of tests and examinations given to the Taiwan aborigines. Preceding it were registration for nativity, anthroposcopic and anthropometric measurements, and PTC taste tests, which together ordinarily took 15 to 20 minutes, thus allowing the subject to rest from walking to the place of examination. Only adults were examined. The subject was seated beside a desk or table about as high as the level of the heart, with the right arm resting on its top. The blood pressure was taken by medical personnel from the County Health Station. Two readings for each person, systolic and diastolic, were taken by the auscultatory method: after a gradual release of air in the armlet the pressure at the first occurrence of sound was taken as the systolic pressure level, and the pressure at the beginning of the fading of sound was taken as the diastolic pressure level. The two readings were recorded in millimeters of mercury for each person. The same instrument was used throughout the survey, and the procedure was kept consistent.

Since age is correlated with blood pressure, variation in mean age between tribes was first examined for the sake of comparability. The number of people tested in each tribe and their mean age are given in Table 32. The means for men ranged from 34.0 in the Tsou to 42.7 in the Ami tribe, and for women from 34.7 in the Puyuma to 41.2 in the Saisiat. Analysis of variance showed that the differences be-

Table 32.   Coefficients of regression of blood pressure on age for people of the Taiwan aboriginal tribes.

		Men					Women		
Tribe	No.	Mean age $\pm$ s.e.	Systolic	Diastolic	No.	Mean age $\pm$ s.e.	Systolic	Diastolic
Atayal	96	25.9 $\pm$ 1.30	−0.05	0.12	146	34.8 $\pm$ 0.95	0.29[a]	0.25[a]
Saisiat	66	40.9 $\pm$ 1.75	.65[b]	.18	60	41.2 $\pm$ 2.00	.74[b]	.27[b]
Bunun	96	39.1 $\pm$ 1.29	.31[b]	.18[a]	110	39.2 $\pm$ 1.31	.41[b]	.20[b]
Tsou	71	34.0 $\pm$ 1.17	− .06	.27	75	35.7 $\pm$ 1.34	.13	.15
Paiwan	126	39.1 $\pm$ 0.94	.45[b]	.16[a]	218	35.5 $\pm$ 0.72	.16	.12[a]
Rukai	111	36.4 $\pm$ 1.13	.44[b]	.40[b]	94	35.3 $\pm$ 1.25	.35[b]	.34[b]
Puyuma	63	35.4 $\pm$ 1.28	.90[b]	.68[b]	110	34.7 $\pm$ 1.00	.27[a]	.30[a]
Ami	145	42.7 $\pm$ 1.15	.04	.03	94	39.4 $\pm$ 1.48	− .03	− .07

[a] $P < 0.05$.   [b] $P < 0.01$.

tween tribes were significant for both men and women. Therefore, age difference should be taken into account in analyzing the blood pressures.

The coefficients of regression on age of the systolic and diastolic pressures for men and women of each tribe were computed separately; the results are given in Table 32. To illustrate the distribution of blood pressures with respect to age a scatter diagram was constructed for each sex and each tribe (Figs. 27, 28, 29, and 30), and a regression line fitted in each graph. The significance of these coefficients is indicated on the table. The coefficients in the Ami and Tsou tribes are not significant, but in the remaining tribes most of them are.

The Saisiat and Puyuma tribes have the highest regression coefficients, the Bunun, Paiwan, and Rukai intermediate, and the Ami and Tsou tribes the lowest. There are some discrepancies between sexes, possibly due to sampling errors, such as the inclusion of a small number of people with profoundly different blood pressures. (These people

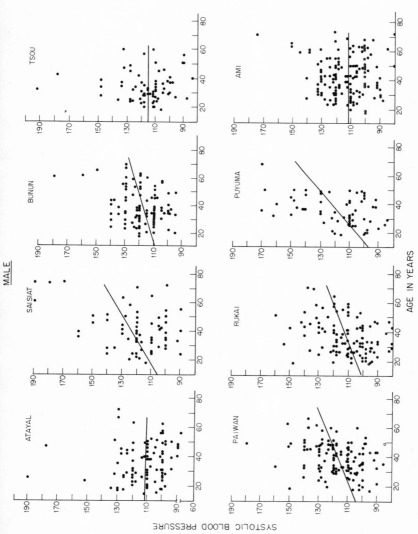

Fig. 27. Scatter diagrams showing the distributions of systolic blood pressures at different ages of men of the different tribes. The regression of blood pressure on age is shown with the fitted regression line in each tribe.

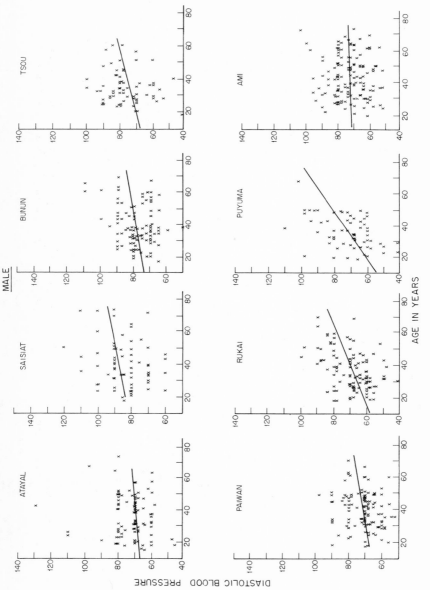

Fig. 28. Scatter diagrams showing the distribution of diastolic blood pressures at different ages of men, constructed as in Fig. 27.

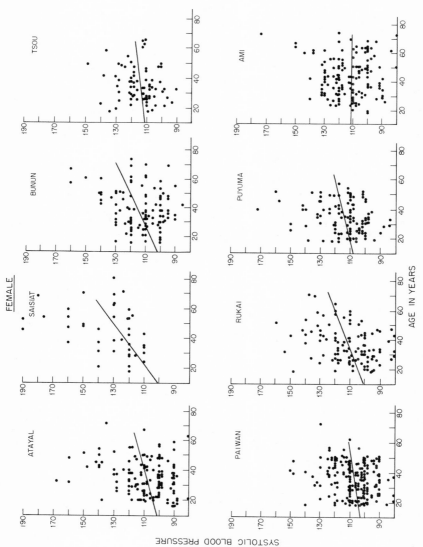

Fig. 29. Scatter diagrams showing the distributions of systolic blood pressures at different ages of women, constructed as in Figure 27.

137

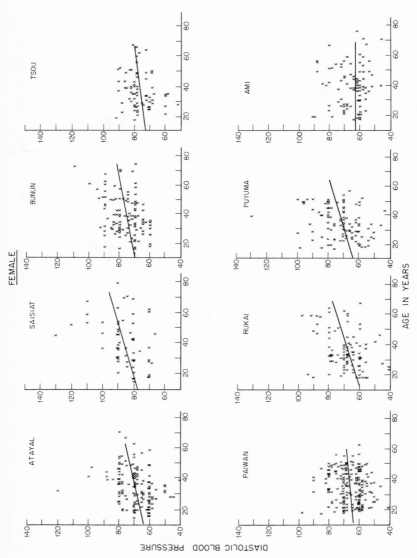

Fig. 30. Scatter diagrams showing the distributions of diastolic blood pressures at different ages of women, constructed as in Fig. 27.

may have had pathologic conditions, but it would have been difficult to select them out from examination.) But in the majority of tribes the regression coefficients are consistent between sexes in respect to difference from zero. This consistency indicates the reliability of the regression estimate, and suggests that intertribal differences are probably real.

Since regression coefficients are significantly different among the tribes (Table 33) the differences between the means of the tribes cannot be analyzed by the covariance method to correct for age difference by a common regression. Rather, the blood pressures of each tribe must be corrected by its own regression. For the sake of calculation, only the means are adjusted to the average age of 38 years, using the regression coefficients of each tribe. The adjusted means, together with the original, are given in Tables 34 and 35. Obviously this correction made little change either in the rank order or the actual values, but to satisfy more statistically minded readers we have used the adjusted means in our comparison.

The mean systolic and diastolic pressures range from 106 and 65 mm Hg in the Ami to 122 and 85 mm Hg in the

Table 33. Test of significance of the difference in regression of the diastolic pressure of women among tribes.

Source of variation	Degree of freedom	Sum of squares	Mean squares	F
Deviation from average regression within tribes	898	99,148.3		
Deviation from individual tribe regression	891	84,478.2	94.8	
Differences among tribe regression	7	14,670.1	2095.7	22.10[a]

[a] $P < 0.01$.

Table 34. Observed and adjusted mean blood pressures (mm Hg) of men of the different aboriginal tribes.

Tribe	Observed[a]		Adjusted	
	Systolic	Diastolic	Systolic	Diastolic
Atayal	111.6 ± 2.30	70.6 ± 1.47	111.5	70.8
Saisiat	122.9 ± 3.19	84.3 ± 1.56	121.0	83.8
Bunun	118.8 ± 1.43	78.5 ± 1.00	118.5	78.3
Tsou	115.8 ± 2.18	74.1 ± 1.76	115.5	75.2
Paiwan	116.8 ± 1.73	71.0 ± 0.83	116.3	71.2
Rukai	111.2 ± 1.59	68.6 ± 1.06	111.8	69.3
Puyuma	119.1 ± 2.75	71.2 ± 1.77	121.5	73.0
Ami	112.9 ± 1.24	73.3 ± 0.91	112.8	73.1
Average	116.1 ± 2.05	74.0 ± 1.30	116.1	74.3

[a] The values following the means are standard errors of the means.

Table 35. Observed and adjusted mean blood pressures (mm Hg) of women of the different aboriginal tribes.

Tribe	Observed[a]		Adjusted	
	Systolic	Diastolic	Systolic	Diastolic
Atayal	108.4 ± 1.34	69.4 ± 1.37	109.3	70.2
Saisiat	124.3 ± 3.13	85.9 ± 1.60	121.9	85.1
Bunun	114.1 ± 1.67	75.4 ± 1.00	113.7	75.1
Tsou	114.6 ± 1.28	76.1 ± 0.99	114.9	76.5
Paiwan	106.9 ± 0.93	66.7 ± 0.57	107.3	67.0
Rukai	106.5 ± 1.37	67.9 ± 1.10	107.5	68.8
Puyuma	114.5 ± 1.47	71.7 ± 1.30	115.5	72.7
Ami	106.7 ± 1.64	64.7 ± 1.14	106.7	64.7
Average	112.0 ± 1.60	72.2 ± 1.13	112.1	72.5

[a] The values following the means are standard errors of the means.

Saisiat tribe. The mean systolic pressure for Saisiat and Puyuma men are highest among the tribes, and lowest in the Ami. The diastolic pressure for men is highest in the Saisiat and lowest in the Rukai. Tests of the differences between the means showed that in general differences of 4.6 mm for systolic and 3.6 mm for diastolic pressure are significant, at least at the 5 percent level. Computation of these values was based on the largest standard deviations of the means of the two tribes and the smallest number of degrees of freedom. This test of significance should be considered conservative. Sex differences in blood pressure within a tribe were not significant in most cases, but the general tendency appears to be that men have higher blood pressure than women, as indicated by the averages for all the tribes. Similarly, the average regression coefficient is greater for men than for women. This differs from what was reported by Bøe et al. (1956).

Based on Hunter's observation (Best and Taylor, 1950) of a quarter million healthy Americans there is an average increase of about 3.6 mm in systolic pressure and 1.0 mm in diastolic pressure for every 10 years of age increase after 20. On this basis the average systolic and diastolic pressures for a 38-year-old American would be about 125 and 83.5, respectively. From the data reported by Bøe et al. (1956) the average for males and females in the age group 35–39 are 132 mm for systolic and 80 mm for diastolic pressure. A study of the blood pressure of some people of Paiwan and Ping Pu was made by Jou and Wang (1960, 1960a). The blood pressure cited by those authors for the Paiwans and Ping Pus at 35 to 39 years of age, together with Chinese at the same ages but in different occupations and habitations, are given in Table 36 for comparison. The average blood pressure of the Paiwans reported by them is comparable to that we obtained for this tribe. The average blood pressure of the Saisiats and Puyumas are comparable with those of Americans and higher than those of some Chinese and the civilized tribe of Ping Pu. Most of the averages of the remaining aboriginal tribes, on the other hand, are lower than Chinese both in systolic and diastolic pressures.

Physical and social environments differ between one society and another, and the differences between primitive tribes and civilized populations are profound. Among the tribes the Saisiat and Puyuma have the highest mean blood pressures and regressions of blood pressure on age, which is even higher than those of civilized populations (Table 35). The distribution for both systolic and diastolic blood pressures of Puyuma women (Fig. 31) is similar to that for the general population shown by Platt (1963). As we have noticed, the Saisiat and the Puyuma are small tribes and most of their people lived in mixed villages with the Chinese people and had more contact, and possibly some free competition, with them. As a factor causing increased blood pressure and regression on age, the foregoing situation is in line with that reported by Cruz-Coke et al. (1964), i.e., the effect of environment changes, from primitive to civilized and competitive, on the age-dependence of blood pressure level of the Easter Islanders, as mentioned in the beginning of this chapter. In connection with the small size of the Saisiat and Puyuma tribes there is the possibility of

Table 36.   Mean blood pressures (mm Hg) of people of different ethnic groups at 35 to 39 years of age (Jou and Wang, 1960, and Jou and Wu, 1960).

Ethnic group	Men			Women		
	No.	Systolic	Diastolic	No.	Systolic	Diastolic
Paiwan	46	110.8	71.1	43	105.7	65.3
Ping Pu	50	115.2	76.4	37	119.3	74.0
Chinese:						
Fishing	75	115.4	68.2	21	110.3	66.0
Insurance	140	117.0	74.1	57	115.6	72.9
Police & railway	131	122.3	69.1			
Taipei inhabitants	244	117.8	76.8	204	118.7	75.7
Average	(590)	118.3	73.4	282	117.4	74.4

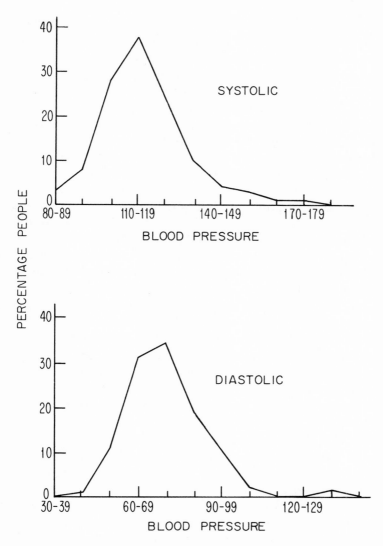

Fig. 31. Diagrams illustrating the distribution of blood pressures of Puyuma women. There is a slight tendency of skewness to the right for the systolic pressure but no obvious tendency of bimodal distribution.

a relatively higher amount of inbreeding in comparison with other tribes. Inbreeding may affect general health and cause early onset of aging, thus affecting blood pressure.

It is interesting that the mean blood pressure was low and that no regression with age occurred in the Ami tribe. It has been claimed that body build is correlated with blood pressure level and that within any weight group broad-chested persons have higher average blood pressures than those of slender build (Robinson and Brucer, 1939), a finding which coincides with the fact that among the tribes the Amis are relatively small and the Bununs relatively large in chest girth (Tables 17 and 18). The mean blood pressure of the Bununs ranks third highest, below that of the Saisiat and Puyuma only. Although body build may be only one factor in blood pressure variations between the tribes, it should be noted that physically the Amis differ most from the other tribes. Whether the difference is mainly a manifestation of a biologic property of the same or different genetic basis is unknown, but the low level of blood pressure together with the lack of regression on age may be considered a physiologic characteristic of the Ami tribe.

As reported by the investigators mentioned in the beginning of this chapter, there are many small ethnic groups of natives or isolates with low blood pressure, some of them also showing less regression on age increase. These groups exist in situations much like those of the aboriginal tribes, especially the Ami. To what extent hypotension is affected by environment—geographic, socio-economic, and climatic —or by genetic background, is difficult to assess. Nevertheless, populations found to have low blood pressure are invariably of Mongoloid origin; so there is a possibility that relatively low blood pressure is characteristic of the genetic predisposition of Mongoloid races.

The high vulnerability of blood pressure to the effect of

environment makes the search for the genetic basis of its variation difficult. This has been mentioned by Osborne et al. (1963). They found by comparing intrapair variance in blood pressure that it was no smaller between monozygotic than between dizygotic twins, and they concluded that the variability in their measured levels was predominantly due to environmental influences. If we consider that essential hypertension is merely the upper end of the range of blood pressure as hypothesized by Pickering (1955)—although there is opposition to this viewpoint (Schweitzer et al. 1962) —the large variation within and between ethnic groups, with respect to both hyper- and hypotension, would be difficult to explain by the single-gene theory even with partial dominance. Furthermore, since blood pressure level is of vital importance to human health, it must be highly adaptive as a genetic character, and a predisposition to high blood pressure would be selected against in both civilized and uncivilized populations. In the case of a single gene its frequency would be low, unless it is balanced by other advantages due to genetic correlations or by a high rate of recurrent mutation. The continuous nature of the distributions and variations between and within populations would be easy to explain by natural selection operating on characters affected by many genes. These genes have effects on other characters due to pleiotropism or linkage, as mentioned in the introduction, and thus their frequencies are maintained in certain levels relative to the internal and external environment of a given population. Such would be also the explanation for the hypotension of the natives in some geographic niche. In regard to the polygenic theory, we are still far from understanding the physical and biologic properties of multiple factors or polygenes. Nevertheless, if the effects of individual polygenes on a given trait are considered different in magnitude, e.g., some genes having stronger effects than others, the hot dispute between the multiple-factorial and single-gene hypotheses may be eased to a great extent.

To summarize, we found low blood pressure levels in most of the Taiwan aborigines and in some tribes the levels are less affected by age than in more civilized populations. Taking into account the reports regarding some other isolates, it would appear that low blood pressure may be a characteristic of genetic predisposition in ethnic groups of Mongoloid origin. There are differences in regression of blood pressure on age and in the adjusted mean blood pressure levels between the tribes.

♀♂

CHAPTER 9. DERMATOGLYPHICS

The history of dermatoglyphics—of fingers, palms, and even soles—is old; exactly how old we do not know, but a fingerprint appears on a Chinese seal made no later than the 3rd century B.C. For a detailed account of some early references see Cummins and Midlo (1961).

Dermatoglyphics as a scientific tool came into use only toward the end of the 19th century. Among the investigators whose work marked milestones in its development was Galton (1895), who pioneered the study of fingerprints in relation to morphology and inheritance, not only bringing dermatoglyphics into the field of science but also actually initiating the methodology of biometrical genetics in his study of fingerprint correlation between relatives. At that time the Mendelian laws of inheritance had not been discovered, and the theory of a multiple factor or polygenic basis for inheritance of quantitative variations was unknown. A few years later Wilder (1902) developed the methodology of plantar and palmar dermatoglyphics as applied to inheritance and racial differences, of which there had been only rudimentary knowledge up to that time. A great advance was then made by Bonnevie (1924), who found in her study of the heritability of fingerprints that correlations of ridge counts between relatives agreed well with their genetic relationships, i.e., 0.92 for monozygotic twins, 0.54 for dizygotic twins, and 0.60 for siblings. She also studied the embryology of finger ridges. More recently Holt (1953) and many others have confirmed Bonnevie's work on the quantitative inheritance of ridge counts and have extended this field. Along with the rapid growth in our knowledge of human genetics in the last 10 to 15 years, many findings have been published on the inheritance of the dermato-glyphics of fingers and palms. Although these studies provide no new discoveries, they have furthered the scientific development of dermatoglyphics, helped establish its basis as a quantitative genetic trait, and proved its usefulness in such applications as racial analysis, diagnosis of twins, etc. The recent wide interest in this field of research is reflected in the Proceedings of the Second International Congress of Human Genetics (Rome) 1961.

Finger and palm prints are unique in that they are quantitative when classified by ridge counts or other metric values, and qualitative when classified by pattern types or similar discrete indices. Though highly heritable according to either classification, no hypothesis has succeeded in basing pattern variation on a single Mendelian gene. Bonnevie (1924), in a three-factor hypothesis for fingerprint pattern variation, postulated that one factor caused thickening of the epidermis as measured by the number of ridges, and that the other two produced a "cushioning" effect which reduced ridge counts in the 2nd and 3rd, and 4th and 5th digits. Grüneberg (1928) proposed a two-factor model for fingerprint pattern variation. Apparently neither of these hypotheses has proved satisfactory.

Development of ridges both in fingers and in palms is completed in the fourth month of pregnancy. Thereafter, environment will not affect them; in number, alignment, and pattern the ridges remain the same throughout life. For this reason they are ideal genetic traits for the study of correlations between relatives, and of differences between individuals or populations, since age as a variable can be disregarded. Furthermore, sexual selection is not involved in the dermatoglyphic characteristics of hands or soles, as it is in some other genetic traits. One cannot, however, eliminate the possibility of linkage with, or pleiotropic effect from, genes subjected to various selection pressures.

Various technics and methods are used for analyzing dermatoglyphic variations—pattern type and intensity, ridge counts, angle of the palm main lines, etc. We are here mainly concerned with ridge patterns, pattern intensity, and number of triradii in the fingers, and with the main ridge lines and patterns in the palms. As most of the customary indices and measurements are highly correlated, we consider a limited number sufficient for studying sexual, bimanual, and tribal variations.

Finger and palm prints of both hands were taken for adults and school children in each tribe. The whole hand, to which black paste had been applied with cheesecloth wrapped around a ball of cotton, was pressed onto a sheet of fairly absorbent paper. The thumb print was made separately, since its ball is oriented differently from that of the other digits and the palm. Printing was repeated for any digit or palm of which the first impression was not clear.

Galton's four major categories were used in classifying the fingerprint types: arch, ulnar loop (opening at the ulnar side), radial loop (opening at the radius side), and whorl; compound patterns were classified as whorl. In coding the patterns for quantitative analysis the scores 0, 1, and 2 were used for arch, loop, and whorl. These codes are based on the number of triradii, which are highly correlated with ridge counts. The loop has one triradius; the whorl, except in compound patterns, two. (Since the arch generally has none, the small number in which triradii appeared were disregarded.)

Palm prints were read according to the method outlined by Cummins and Midlo (1961). We used only the ridge patterns of the thenar and hypothenar eminences, and the main lines from the four digital triradii located in proximal relation to the bases of digits 2, 3, 4, and 5; in the radio-ulnar sequence these lines are designated A, B, C, and D. Two distal radiants embrace the base of the digits and the proximal radiants (the main lines) travel across the interior of the palm and terminate at various points on the periphery of the palm. Numbers 1 to 13, which designate the regions of termination of the palmar main lines, are the index numbers for the lines. A zero score is given to the aborted main line (see Cummins and Midlo for descriptions of "abortion"). The thenar eminence is the elevation in the proximoradial quadrangle of the palm, and the hypothenar eminence is that lying in the ulnar portion of the palm. The same classification is used for ridge patterns in these regions as for those in the finger balls. For quantitative analysis these indices for the main line readings are used directly, and the same scores for the fingerprint ridge patterns are used for those in the thenar and hypothenar regions. A diagram (Fig. 32, left) of fingerprint types, palm triradii, the main lines, and thenar and hypothenar ridge patterns is given for readers not familiar with dermatoglyphics, together with a finger and palm print (Fig. 32, right) made during our survey.

Only complete prints of both hands were used. When the data were scrutinized for analysis many prints were discarded, either because hand defects had made complete prints impossible to obtain, or because the blurred ridges and partly sloughed dermis of old age or malnutrition had made them unreadable.

FINGERPRINTS

Finger Differences

Ulnar loops and whorls show the largest frequencies in each digit of both hands of both sexes (Fig. 32). Arches and radial loops occur more frequently in the second digit than in any other, but are completely absent in some digits in certain tribes. In pattern distribution, second digits are most variable, and fifth digits, with the generally highest frequencies of ulnar loops, is least variable; the other digits are intermediate. These variabilities are consistent both for hands and for sexes. The relative variability in the fingerprint pattern among the fingers may be associated with an evolutionary effect of relative usage and versatility in performance: i.e., fingers used more frequently and performing more different functions may have developed greater variability in fingerprint patterns. When pattern frequencies were pooled for the five digits of each hand, the percentages of ulnar loops were highest for most tribes.

Hand and Sexual Differences

The different patterns vary in frequency between left and right hands. Tables 37 and 38 show that the frequency of whorls and radial loops is higher in men, and higher in the right hand than in the left. These differences are consistent throughout the tribes except that in the Puyuma the frequency of the radial loop is slightly greater in women than in men. In women there are more arches in the left hand than in the right, but differences in the other patterns vary depending on the tribes. The significance of the differences between left and right hands was analyzed by chi-square statistics, based on the pattern frequencies in each digit of each hand, pooling the samples of all the tribes but keeping sexes separate (Table 39). The total chi square for men is more than twice that for women. Ulnar loop and whorl contribute the highest chi squares for men, and the arch for women.

For analysis of the differences between sexes we also used chi-square tests, since the sexual percentages for each digit

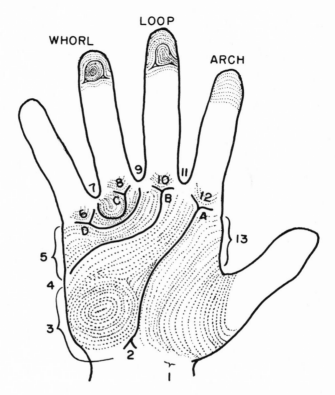

Fig. 32 (*left*). Diagram illustrating the fingerprint types, the main ridge lines, and the thenar and hypothenar ridge patterns. The fingerprint patterns are self-explanatory. In the palm, A, B, C, and D are the four main lines. The numbers from 1 to 13 mark the peripheral regions of the palm where the main lines terminate. In this case the main line formula is 2, 4, 7, 9. There is no true pattern in the thenar, but there is a whorl in the hypothenar region.

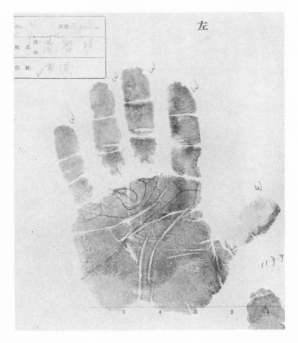

Fig. 32 (*right*). The left handprint of a chieftain of the Puyuma tribe. His fingerprint patterns from the 1st to the 5th digit are all whorls. The palm formula is 3, 7, 7, 11. There is no true pattern in the thenar or hypothenar regions. In the upper left-hand corner is the identification of the subject. The letter W over each finger means "whorl," the reading of the fingerprint. The numerals at the bottom are for reprinting individual fingers in case the prints above are not clear; in this case only the thumb was reprinted.

Table 37. Frequency distributions of fingerprint patterns and values of χ^2 fo

| | | Left hand | | | |
| | | Loop | | | Sub-total |
Tribe (number)	Arch	Radial	Ulnar	Whorl	
Atayal (92)					
Freq. obs.	8	12	231	209	
χ^2	1.21	0.13	0.08	0.45	1.87
Saisiat (99)					
Freq. obs.	17	17	293	168	
χ^2	1.48	0.51	6.24	10.15	18.38
Bunun (141)					
Freq. obs.	9	22	398	276	
χ^2	4.54	0.13	3.86	2.90	11.42
Tsou (99)					
Freq. obs.	9	15	315	156	
χ^2	1.06	0.03	15.06	16.04	32.20
Paiwan (166)					
Freq. obs.	28	22	459	321	
χ^2	2.14	0.17	2.78	4.22	9.31
Rukai (102)					
Freq. obs.	20	13	228	249	
χ^2	3.69	0.21	4.15	3.50	11.54
Puyuma (65)					
Freq. obs.	15	14	135	161	
χ^2	5.36	2.26	5.88	2.85	16.36
Ami (142)					
Freq. obs.	10	16	259	425	
χ^2	3.68	1.00	0.11	44.51	49.29
Total (906)					
Freq. obs.	116	131	2318	1965	
Percent	3	3	51	43	
χ^2	23.17	4.42	38.15	84.61	150.36

men of the Taiwan aboriginal tribes.

| | Right hand | | | | |
| Arch | Loop | | Whorl | Sub-total | Total |
	Radial	Ulnar			
2	17	226	215		
5.55	0.20	0.89	0.09	6.73	8.60
16	21	271	187		
3.92	0.02	7.93	10.25	22.12	40.50
3	37	327	338		
8.60	2.18	0.01	0.01	10.79	22.22
7	28	264	196		
0.80	2.84	5.54	6.84	16.02	48.22
24	43	406	357		
3.49	2.26	1.38	3.86	10.98	20.29
16	9	216	269		
3.45	6.87	1.95	2.70	14.96	26.51
13	16	141	155		
6.68	0.51	0.54	0.00	7.73	24.09
9	16	240	445		
1.82	6.00	23.46	33.28	64.56	113.85
90	187	2091	2162		
2	4	46	48		
34.31	20.87	41.68	57.03	153.89	

Table 38.   Frequency distributions of fingerprint patterns and values of χ^2 fo

Tribe (number)	Arch	Loop Radial	Loop Ulnar	Whorl	Sub-total
		Left hand			
Atayal (145)					
Freq. obs.	15	18	379	313	
χ^2	2.31	0.08	0.20	1.01	3.60
Saisiat (97)					
Freq. obs.	13	22	282	168	
χ^2	0.23	6.42	2.04	4.57	13.25
Bunun (163)					
Freq. obs.	9	24	478	304	
χ^2	10.18	0.25	4.19	2.50	17.12
Tsou (92)					
Freq. obs.	28	8	277	147	
χ^2	13.76	1.47	3.99	8.89	28.11
Paiwan (256)					
Freq. obs.	56	33	765	426	
χ^2	7.23	0.03	9.68	17.91	34.85
Rukai (80)					
Freq. obs.	11	9	186	194	
χ^2	0.13	0.25	3.58	5.75	9.70
Puyuma (90)					
Freq. obs.	20	10	187	233	
χ^2	2.82	0.32	11.84	13.18	28.17
Ami (81)					
Freq. obs.	2	10	128	265	
χ^2	8.71	0.06	36.05	59.98	104.80
Total (1004)					
Freq. obs.	154	134	2682	2050	
Percent	3	3	53	41	
χ^2	45.36	8.88	71.56	113.78	239.58

women of the Taiwan aboriginal tribes.

	Right hand				
	Loop			Sub-	
Arch	Radial	Ulnar	Whorl	total	Total
6	18	400	301		
5.15	0.48	0.02	0.27	5.92	9.52
10	13	280	182		
0.00	0.75	0.43	0.95	2.14	15.39
5	19	456	335		
8.05	0.19	0.03	0.12	8.39	25.50
13	13	276	158		
1.43	1.11	1.68	4.09	8.32	36.43
37	20	815	408		
4.67	1.82	15.40	22.74	44.63	79.48
12	4	207	177		
1.85	2.34	0.00	1.52	5.70	15.40
16	10	215	209		
5.15	0.03	4.84	4.16	14.17	42.33
3	9	138	255		
3.31	0.02	33.50	51.38	88.22	193.01
102	106	2787	2025		
2	2	56	40		
29.62	6.74	55.90	85.22	177.48	

Table 39.   Chi-square tests of bimanual and sexual differences in frequencies of occurrence of fingerprint types, pooling samples from all the tribes.

Sex	Hand	Frequency	Arch	Loop		Whorl	Total
				Ulnar	Radial		
Male	Left	Observed	116	2318	131	1965	4530
		Percent	2.6	51.2	2.9	43.4	
	Right	Observed	90	2091	187	2162	4530
		Percent	2.0	46.2	4.1	47.7	
		χ^2	35.00	11.69	9.86	9.40	65.95[b]
Female	Left	Observed	154	2682	134	2050	5020
		Percent	3.1	53.4	2.7	40.8	
	Right	Observed	102	2787	106	2025	5020
		Percent	2.0	55.5	2.1	40.3	
		χ^2	10.56	2.02	3.27	0.15	16.00[b]
Sex difference							
	Left	χ^2	2.17	2.31	0.42	3.66	8.55[a]
	Right	χ^2	0.12	40.83	31.51	29.64	102.10[b]

[a] Significant above 5 percent level.   [b] Significant above 1 percent level.

correspond to those for right and left hand differences; the chi-square values are given in the last two lines of Table 39 and the observed frequencies are omitted. The total chi squares for sex differences are significant for both hands, but about 11 times greater for the right hand than for the left.

Assuming no sexual selection for dermatoglyphic characteristics and their development, sexual differences in dermatoglyphics may be of evolutionary significance, the variations resulting from different use of the hands by men and women and differences in rates of growth and development. This explanation can be similarly applied to bimanual differences, since the right hand often performs differently from the left. The genetics of asymmetry has recently been of considerable interest to the Drosophila geneticists (Reeve, 1960). In studies of the chaeta number in Drosophila, asymmetry is considered "developmental noise" (Waddington et al., 1957). Perhaps that is true when it occurs randomly in individuals within a population. If the asymmetry is consistent between one hand and the other in a population or populations, there may be some underlying evolutionary basis.

Tribal Differences (Figs. 33 and 34)

Digit 1. Relatively large variations appear among the tribes in the proportion of ulnar loops and whorls in this digit. When hands and sexes are combined, the percentage of whorls is greater than that of ulnar loops in the Rukai, Puyuma, and Ami tribes and consistent for hands and sexes. In the remaining tribes the relative proportion of these two patterns varies between hands and between sexes. The percentage of whorls appears to be lowest in the Saisiat and Tsou tribes.

Digit 2. In general whorls occur here most frequently in all tribes, with ulnar loops next and arches least. This order is consistent for both hands and sexes in the Ami, Atayal, and Bunun tribes. The percentage of ulnar loops is greater than that of whorls in both hands of females and in the left hands of Paiwan males.

Digit 3. The percentage of ulnar loops is greater than that of whorls in each tribe except Ami, where the reverse is true. This is consistent for both hands and sexes.

Digit 4. The relative proportion of ulnar loops and whorls varies among the tribes, but in general whorls outnumber ulnar loops. The Ami has the highest percentage of whorls in this digit. Differences in the percentages of ulnar loops

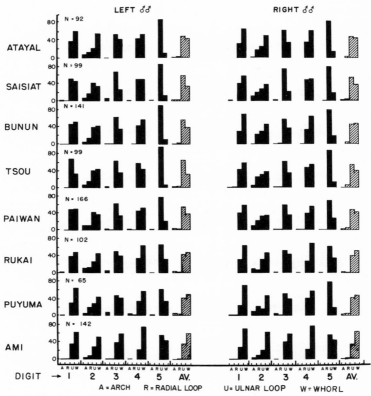

Fig. 33. Percentage frequency distribution of fingerprint patterns in men of the different tribes. The left half is for the left hands and the right half for the right. The solid black histograms are for individual digits, and the hatched histograms are the average percentages of each pattern for the five digits.

and whorls are rather small in the Saisiat, Bunun, Tsou, and Paiwan compared to the other tribes.

Digit 5. The percentage of ulnar loops is greater than that of whorls for both hands and both sexes throughout the

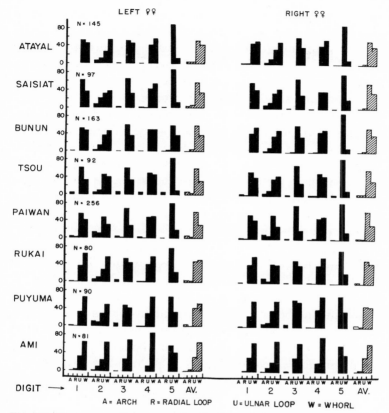

Fig. 34. Percentage frequency distribution of fingerprint patterns in women of the different tribes, constructed like Fig. 33.

tribes, but again the percentage of whorls is greater in the Ami than in the others.

When the fingerprint pattern frequencies for the five digits are pooled, the Ami stands out from the remaining tribes for its consistently greater percentage of whorls over ulnar loops; the Atayal, Saisiat, Tsou, and Paiwan tribes have a greater percentage of ulnar loops than of whorls; these percentages are consistent for both hands and sexes. In the remaining tribes the relative proportion between ulnar loops and whorls varies between hands and between sexes. The arches and radial loops occur in rather low frequencies and are difficult to compare.

A test of the differences in the frequencies of each pattern was made among the eight tribes by the chi-square method. Since there are differences between sexes and between right and left hands, separate analyses were made. Expected frequencies used for the test were the averages for the eight tribes. Therefore the total chi squares for the eight tribes are a test of heterogeneity, and the chi square for an individual tribe is the contribution of that tribe to the total chi square. Thus the larger the chi square for a tribe, the greater is its difference from the other tribes as a whole. The results of the tests for tribal differences are given in Tables 37 and 38. For 21 degrees of freedom the total chi square for each hand is significant beyond the 1 percent probability level. The last line of the table shows that the whorl contributes most to the chi square for each hand, or about 50 percent of the total chi square, except that the right hand of men contributes about one-third of the total. The next largest chi-square value is from the ulnar loop and the next from the arch; these chi squares are highly significant. The radial loop contributes least to the chi squares and is significant only in the right hand of men.

The four chi-square totals (one for each hand of each sex), are all significant. The chi square for the right hand of men is greater than that for women. The two chi-square totals, one for men and one for women, are also significant; but the chi square for men is greater than that for women.

The subtotal chi square for each tribe reveals that the largest chi square for each hand of each sex is in the Ami tribe. For both hands of men the second largest chi square is in the Paiwan tribe, the next in the Saisiat, and the least in the Atayal tribe. For both hands of women the second largest chi square is in the Paiwan tribe, the next in the Puyuma, and the least again in the Atayal. Except for the Atayal tribe all chi squares for either men or women of each tribe are significant for six degrees of freedom. As the chi-square test measures the deviation of observed from calculated frequency, based on the percentages of each fingerprint pattern in all the tribes, the chi-square values for any entry actually represent the difference between that tribe and the respective average frequencies of all the tribes. For any two tribes the chi-square value tells only the relative magnitude of the deviation from the mean of the total population, which, of course, can be either in the minus or plus direction.

Since the whorl and loop occur in high percentages in each tribe, the ratio of number of whorls to number of loops may be used for a general comparison of the relationship between tribes at this stage. The ratios for the eastern tribes appear greater than those for the western, excepting the Rukai (Table 40). It should be pointed out here that

Table 40.   Whorl-to-loop ratios in the fingerprints of people of the different tribes, grouped by hand and sex, and both hands and both sexes.

Tribe	Females			Males			Total
	Left	Right	Total	Left	Right	Total	
Atayal	0.79	0.72	0.76	0.86	0.88	0.87	0.80
Saisiat	.55	.62	.59	.54	.64	.59	.59
Bunun	.61	.73	.65	.66	.93	.78	.71
Tsou	.52	.55	.53	.47	.67	.57	.55
Paiwan	.53	.49	.51	.67	.80	.73	.59
Rukai	.99	.84	.91	1.03	1.20	1.11	1.02
Puyuma	1.18	.93	1.05	1.08	0.99	1.03	1.04
Ami	1.92	1.73	1.82	1.55	1.74	1.64	1.70

though the Rukai is considered a western tribe, a good portion of its people live on the east side of Tawu Mountain, in close contact with the Puyumas and to some extent with the Amis. A portion of our sample was taken from the eastern area, and the large whorl and loop ratio obtained may indicate an interflow of genes from the eastern tribes to the Rukai.

Except for minor discrepancies, the general results showing fingerprint pattern variation agree well with the results of the analyses of the physical characters.

Pattern Intensity

Pattern intensity of the finger ridges per individual was computed for each tribe according to the method of Cummins and Staggerda (1935), based essentially on the number of triradii per person. The score was zero for arch, one for loop, and two for whorl. As in scoring the patterns for quantitative analysis, we ignored the rare circumstance where an arch other than a plain arch had a triradius. Pattern intensity of finger ridges is highly correlated with ridge counts. The results indicate that the Ami tribe is distinct from the others. The average intensities for the different tribes are given in Table 41, listed according to hand and sex and averaged for both sexes. There is a great contrast between men and women in the bimanual differences: men have greater pattern intensity in the right hand than in the left; women do not. Differences between sexes vary; the value for one sex is not consistently greater than that for the other in all the tribes, based either on the same hand or on both together. The averages of both hands for the tribes show that the Ami has the largest average intensity, 16.2, far above those for the other tribes of which the average intensities range from 13.2 for the Tsou to 14.5 for the Rukai and Puyuma tribes. These results again indicate that

Table 41. Average finger ridge pattern intensities of the Taiwan aboriginal tribes, listed according to hand, sex, and averages for both sexes.

Tribe	Male			Female			Average
	Left	Right	Both	Left	Right	Both	
Atayal	7.18	7.32	14.50	7.06	7.03	14.09	14.30
Saisiat	6.53	6.72	13.25	6.60	6.77	13.37	13.31
Bunun	6.89	7.38	14.27	6.81	7.02	13.83	14.05
Tsou	6.48	6.91	13.39	6.29	6.58	12.87	13.13
Paiwan	6.77	7.01	13.78	6.45	6.45	12.90	13.34
Rukai	7.26	7.48	14.74	7.28	7.06	14.34	14.54
Puyuma	7.25	7.18	14.43	7.37	7.14	14.51	14.47
Ami	7.92	8.07	15.99	8.25	8.11	16.36	16.18
Average	7.04	7.26	14.29	7.01	7.02	14.03	14.16

the eastern tribes are genetically different from the western tribes. It may be necessary to point out that pattern intensity is highly correlated with ridge count. The relative order in pattern intensity among the tribes was predictably similar to that based on the total ridge count, when a count was taken.

PALMAR MAIN LINES

Main Line Differences (Figs. 35 and 36 and Table 42)

The A line. The indices of the A line range from 1 to 5, with some very small percentages at 7 and 11. The modes fall at 3 in the left hands and at 5 in the right of each sex of each tribe.

The B line. B line indices range in general from 5 to 7, with rather low percentages at 0, 1, 3, 8, and 9 in hands of certain sexes and tribes. The mode is at 5 for all tribes and is the largest among the four main lines. The distribution of the indices is least variable among the lines.

The C line. The 0 index is highest in the C line and that

is one of the characteristics of this line. Other frequently occurring indices are 5, 6, 7, and 9. The index 8 is extremely rare for the reason that the number 8 region is the base of the C line and it is difficult for the C line to travel back to the same region. The mode varies among hands, sexes, and tribes. It occurs at either 5, 6, or 7 in most cases except in the right hands of men, where it is at 6 for practically all the tribes.

The D line. In the D line the indices range from 7 to 11 in general, with some extremely low frequencies at 0 and 1. The mode varies between hands, sexes, and tribes; in general it is at 7 in the left hands, and in the right hands it is at either 9 or 11, for all the tribes.

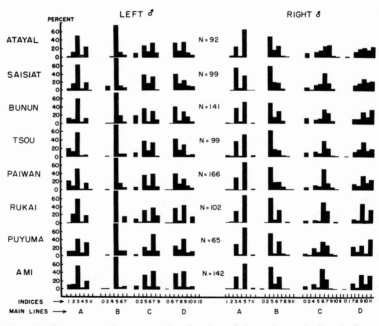

Fig. 35. Percentage frequency distribution of the palm main line indices for the men of each tribe. For description of the main lines and indices see Fig. 32 and the text.

Hand and Sexual Differences

It is clear from the diagrams (Figs. 35 and 36) that there are great differences between hands in the percentage distribution. In both sexes the mode for the A line is at the 3 position in the left hand and at the 5 in the right, and there is more variation in the left than in the right for practically all the tribes. In the B line the modes are the same for both hands, but the percentages at the 7 index are greater, and at 5 are smaller in the right hand than in the left; this is consistent throughout the tribes. The magnitude of variation in the C line is not obvious between left and right hands, but the modes in the left hand are at the 5 and 7

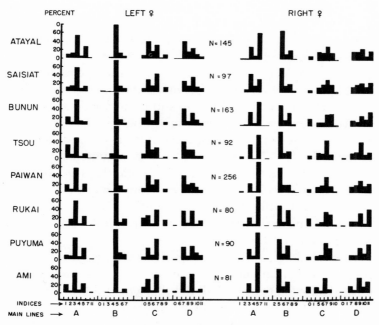

Fig. 36. Percentage frequency distribution of the palm main line indices for the women of each tribe, constructed like Fig. 35.

positions and in the right are mostly at the 6 position in men and at 7 in women. Differences are shown in the D line; the variation in the modes and distributions between hands are interesting. The modes are at the 7 or 9 position in the left hands, and at 9 or 11 in the right. The distributions are skewed to the left in the right hands, and to the right in the left hands; thus the left- and right-hand distributions form almost mirror images to each other. It is quite clear that distributions for each main line tend toward greater values in the right hand than in the left, in both sexes.

A chi-square test was made between the left and the right hands in each sex, pooling the frequencies of all the tribes together (Table 42). Since sex differences are rather consistent throughout the tribes, the pooled analysis should introduce little bias while saving a great deal of calculation. The chi square is significant between hands in each sex, and consequently the total chi square is also significant.

Sex differences in the main line indices are less than the hand differences. In general, the men tend to have the higher indices for the right hand main lines. The differences were analyzed by the chi-square method, pooling all the tribes together but keeping the left and right hands separate. The results are summarized in the last two columns of Table 43. The total chi square is significant, the A line contributing the most and the B line the least. But the total chi square for sex differences is much smaller than that for bimanual differences. The evolutionary significance of bimanual and sexual differences as suggested for fingerprints should hold also for the main lines.

In analyzing sex and hand differences for either fingerprints or main line indices a chi-square test should be performed for each tribe and all the chi squares added together. This would involve rather heavy calculations, however. We are aware that pooling the tribes may reduce differences, if they are in opposite directions, and that weights for the different tribes could vary according to sample size; but as a crude test, perhaps also a conservative one for the above reason, the pooled analysis is considered sufficient for the present purpose.

Average Indices

Average indices for the main lines in a tribe represent essentially the transversary of the alignment of the main lines, i.e., the higher the index is, the more the transversary. The contrast between left and right hands is more easily visualized from these values (Table 43), as shown by the consistent difference of larger indices of the right over the left hand for both men and women throughout practically all the tribes. But no consistent difference between sexes is shown; the average for men is greater in some tribes and less in others. When the two sexes are averaged, tribal differences are rather small compared to bimanual differences. The Tsou tribe has the smallest average index, 5.96, and the Rukai the largest, 6.40. Nevertheless, these results also show some separation between the eastern and western tribes except the Rukai. Although the Rukai is considered a western tribe, some of its people inhabit the east side of Tawu mountain, as previously mentioned. The results of the main line indices are generally in agreement with the fingerprints and even with physical measurements.

Thenar and hypothenar patterns are in general rare in all the tribes, and analyses of the data for percentage differences among the tribes, between sexes, and between hands are omitted.

Tribal Differences (Figs. 35 and 36)

The Saisiat tribe is unique at the A line, showing the mode at 3 instead of at 5 as in the other tribes in the right

Table 42. Chi-square tests of bimanual and sexual differences in the frequencies of main line indices, pooling samples from all the tribes.

| Main line | Region | Female | | | | | Male | | | | | Sex differences | |
| | | Left | | Right | | | Left | | Right | | | Left | Right |
		Freq. obs.	%	Freq. obs.	%	χ^2	Freq. obs.	%	Freq. obs.	%	χ^2	χ^2	χ^2
A	0–2	226	22.5	33	3.3	143.8	220	24.3	27	3.0	151.2	0.6	0.2
	3	547	54.5	280	27.9	86.2	502	55.4	309	34.1	45.9	0.1	6.0
	4	49	4.9	56	5.6	0.5	32	3.5	18	2.0	3.9	2.1	15.1
	5–11	182	18.1	635	63.3	251.1	152	16.8	552	60.9	227.6	0.5	0.4
Subtotal		1004		1004		481.6	906		906		428.5	3.3	21.7
B	3–5	805	80.2	580	57.8	36.5	747	82.5	502	55.4	48.1	0.3	0.5
	6	119	11.9	144	14.3	2.4	93	10.3	133	14.7	7.1	1.1	0.0
	7–9	80	8.0	280	27.9	111.3	66	7.3	271	29.9	124.7	0.3	0.7
Subtotal		1004		1004		150.2	906		906		179.8	1.7	1.2

C	4-5	453	45.1	245	24.4	61.9	400	44.2	188	20.8	76.6	0.1	2.8
	6	125	12.5	144	14.3	1.3	89	9.8	93	10.3	0.1	3.0	6.4
	7-8	350	34.9	367	36.6	0.4	341	37.6	377	41.6	1.8	1.0	3.1
	9	76	7.6	248	24.7	91.5	76	8.4	248	27.4	91.2	0.4	0.8
Subtotal		1004		1004		155.2	906		906		169.7	4.5	13.1
D	0-7	392	39.0	152	15.1	105.1	354	39.1	121	13.4	114.3	0.0	1.1
	8	141	14.0	148	14.7	0.2	102	11.3	100	11.0	0.0	2.9	5.0
	9	292	29.1	294	29.3	0.0	304	33.6	294	32.5	0.2	3.0	1.6
	10	108	10.8	138	13.7	3.7	80	8.8	129	14.2	11.5	1.8	0.1
	11	71	7.1	272	27.1	117.7	66	7.3	262	28.9	117.1	0.0	0.6
Subtotal		1004		1004		226.6	906		906		243.2	7.8	8.3
Total		4016		4016		1013.6	3624		3624		1021.2	17.2	44.3

Table 43. Average main line indices for the Taiwan aboriginal tribes, listed according to sex and hand, and averages for both hands and sexes.

Tribe	Male			Female			Average
	Left	Right	Average	Left	Right	Average	
Atayal	5.86	6.69	6.28	5.74	6.45	6.10	6.19
Saisiat	5.68	6.44	6.06	5.83	6.41	6.12	6.09
Bunun	5.70	6.76	6.23	5.65	6.78	6.22	6.23
Tsou	5.49	6.52	6.01	5.45	6.35	5.90	5.96
Paiwan	5.66	6.70	6.18	5.66	6.55	6.11	6.15
Rukai	5.90	6.78	6.34	6.02	6.90	6.46	6.40
Puyuma	5.98	6.73	6.36	5.98	6.85	6.42	6.39
Ami	5.71	6.74	6.23	5.76	6.76	6.26	6.25
Average	5.75	6.67	6.21	5.76	6.63	6.20	6.21

hands of both sexes. In the B line the mode falls at 5 for all the tribes, and there is not much variation between tribes in the general pattern of the distribution; a small percentage at index 3 occurs in the left hands of both sexes in the Saisiats; this is higher than in any of the other tribes, in most of which it is at zero. Variation in the C line of right hands is also comparatively small among the tribes; the mode falls at index 7 in each, except in the Atayal, where the percentage at index 9 is slightly higher. However, in the left hand the mode varies among the tribes, falling either at 5 or at 7. The C line with zero index does not occur in the left hands of Saisiat males, whereas it does occur in both hands and sexes in all the other tribes. In the D line, a considerable variation in distribution is shown; the mode, falling either at 7 or at 9 in the left hands, and at 9 or 11 in the right, varies among the tribes. There appear to be much higher percentages at the indices 7 and 9 in the left hands, and 9 and 11 in the right hands, of the Rukai, Puyuma, and Ami than of the other tribes.

The percentage distributions of the main line indices show that with a single exception there is no clear-cut separation or grouping of the tribes. The Saisiat tribe does appear to vary from the others at the A line of the right hand in both sexes. The Saisiat is smallest of the tribes, and a relatively large percentage of inbreeding can be expected, but the shift of modes of indices of the A line is difficult to interpret except as a characteristic of the Saisiat tribe.

A chi-square test was made of differences among the tribes for each hand and each sex. Chi squares were computed for the indices of each main line as they were for the fingerprint patterns. The expected frequencies were based on the average frequencies of all the tribes. The tables of the individual chi squares are too extensive to include here but a summary is given for the chi squares for each hand and each sex (Table 44). The chi squares are all significant beyond a 1 percent probability level except that for the Atayal, which is significant at the 5 percent level. But it is quite obvious that the chi squares for the Saisiat, Tsou, and Rukai tribes are much greater than those for the others. These results are in contrast to those of the fingerprint patterns in which the chi square for the Ami is the largest. This discrepancy means that a great difference exists in the percentages of frequencies of the Ami from the tribal populations as a whole in the fingerprint patterns, but not in the main line indices. This may indicate that there are possibly two sets of genes, one more concerned with the fingerprint patterns and the other with the courses of the main line distribution. The gene frequencies in the Ami tribe for the former differ more than in the latter from those of the averages for the tribal populations.

Table 44. Total χ^2 for the difference between the Taiwan aboriginal tribes in the index regions of each of the palmar main lines.

Tribe	Male				Female				Totals	
	No. of people	Left	Right	Sub-total	No. of people	Left	Right	Sub-total	No. of people	χ^2
Atayal	92	19.30	23.41	42.71	145	9.50	20.43	29.93	237	72.6[a]
Saisiat	99	21.95	41.79	63.74	97	19.21	55.14	74.35	196	138.1[b]
Bunun	141	42.33	14.04	56.37	163	23.42	19.84	43.26	304	99.6[b]
Tsou	99	32.44	22.24	54.68	92	40.81	24.13	64.94	191	119.6[b]
Paiwan	166	31.98	15.66	47.64	256	20.36	25.53	45.89	422	93.5[b]
Rukai	102	67.73	35.48	103.21	80	38.34	22.57	60.91	182	164.1[b]
Puyuma	65	27.52	26.45	53.97	90	26.74	14.80	41.54	155	95.5[b]
Ami	142	20.21	16.48	36.69	81	42.84	16.40	59.24	223	95.9[b]
Totals	906	263.46	195.55	459.01	1004	221.22	198.84	420.06	1910	879.1[b]

[a] Significant above 5 percent level. [b] Significant above 1 percent level.

CORRELATION ANALYSIS

A correlation analysis was made of the ridge patterns between digits of the same hand and between the left and right hands of the subject. A similar analysis was made of the main line indices. Only the Paiwan data were analyzed, since the sample from this tribe was the largest. The least-squares method of computing correlation coefficients was used. Values used for the analysis were the codes for fingerprint patterns and the indices for the main lines.

The correlation coefficients are arranged in matrix form in Tables 45 (fingerprints) and 46 (main lines). In each table the upper right triangle matrix is for men and the lower left for women. Each triangle matrix contains two small triangle matrices for the coefficients between fingerprints or main lines of the same hands, and one square matrix for the coefficients between fingers or main lines of different hands.

Table 45. Correlation coefficients of print patterns between fingers of the same and different hands of people of the Paiwan tribe.

		Right					Left				
		1	2	3	4	5	1	2	3	4	5
						Male					
R	1		0.46	0.47	0.37	0.13	0.41	0.36	0.37	0.34	0.18
	2	0.39		.53	.42	.27	.37	.67	.47	.43	.26
	3	.40	.52		.55	.31	.41	.51	.67	.47	.32
	4	.30	.38	.50		.28	.24	.52	.54	.62	.32
	5	.30	.35	.41	.38		.21	.28	.21	.24	.53
L	1	.57	.49	.42	.37	.33		.32	.35	.36	.22
	2	.32	.68	.51	.48	.37	.47		.51	.60	.31
	3	.35	.55	.66	.49	.35	.44	.55		.64	.34
	4	.30	.42	.52	.68	.39	.40	.54	.53		.31
	5	.30	.37	.45	.40	.64	.30	.41	.40	.39	
						Female					

Table 46. Correlation coefficients of the indices between different main lines in the same and different hands of people of the Paiwan tribe.

		Right				Left			
		D	C	B	A	D	C	B	A
					Male				
R	D		0.29	0.86	0.39	0.51	0.17	0.35	0.29
	C	0.37		.23	.01	.10	.52	−.04	−.11
	B	.71	.42		.36	.53	.22	.44	.31
	A	.25	.07	.26		.25	−.02	.24	.43
L	D	.45	.25	.53	.32		.25	.63	.25
	C	.13	.41	.21	.06	.31		−.05	−.07
	B	.35	.39	.55	.25	.69	.21		.29
	A	.08	.06	.15	.35	.21	−.08	.24	
					Female				

For the fingerprints of the same hand, the largest correlation coefficients appear between the 2nd and 3rd, 2nd and 4th, and 3rd and 4th digits, and the least between the 1st and 5th. There is a tendency toward greater correlation for adjacent digits than for those far apart. Also, the correlation values of the left hand are greater than those of the right for both sexes. The correlation coefficients between the digits of right and left hands are shown in the 5 × 5 square matrices at the upper right and lower left corners of Table 45. The values on the main diagonal of each are between homologous digits, and are measurements of the symmetry of individual hands. These values are almost all higher than those for nonhomologous digits, indicating some degree of symmetry in the fingerprint patterns for all. Among them the 2nd, 3rd, and 4th digits are more symmetrical than the 1st and 5th. There is a tendency for greater correlations between digits more adjacent to the homologous digit. In general the correlations for women are greater than those for men. The positive correlations between digits indicate the likelihood of occurrence of fingerprints with similar numbers of triradii. For instance, if a whorl is observed in one digit it is likely that whorls will be found in the others, and also perhaps loops; but arches will be unlikely.

The correlations of the indices between the main ridge lines of the palm are presented in Table 46. They are more variable than those of the fingerprint patterns. In the same hands the correlation coefficients between the B and D lines are the greatest, and those between the A and C are close to zero in both hands of both sexes. The others, between C and D, A and D, and A and B, are less than those between B and D. The correlation between B and C is not significant in the left hands of men, but is much higher and significant in the right hands of men and in both hands of women. The correlation coefficients for the main lines between hands are given in the 4 × 4 square matrix at the upper right and lower left corners of the table. As for the fingerprints, the correlations on the main diagonal are in general higher than the values off the diagonal, except that between B and D, which is higher than many values on the main diagonal. Comparing these correlation coefficients with those for the fingerprints between hands, the main lines are less symmetrical than fingerprint patterns. Some of the other correlation coefficients between hands are small and not significant. Cummins and Midlo (1961) reported an average correlation coefficient between the same main lines in right and left hands of 0.55, which is comparable with, although slightly higher than, those obtained here.

Concerning the genetics of dermatoglyphics, Cummins and Midlo pointed out that ridge alignment is conditioned by the stresses and tension incident to the general growth of the part. The alignments of these ridges in response to the stresses of growth are as the ridge alignments of sand exposed to wind and wave. This view was to some extent shared by Stern (1949). Their interpretation as I understand it implies that the formation of ridge patterns is subject to environmental influences although they are fixed at the early embryonic stage. My opinion is that this hypothesis should not be assumed to the extent of ruling out genetic effect, since it is a fact that ridge counts show very high heritability, and that correlations do exist between ridge patterns and between main line indices. Furthermore, the evidence of greater correlation between adjacent digits than between digits farther apart suggests the existence of a common genetic agent or agents controlling the basic growth pattern of the derms in a region which possibly covers two or more adjacent digits. But the genetic mechanisms remain to be discovered.

DISCRIMINANT ANALYSIS

Discriminant analysis of the dermatoglyphic data was made in the same way as that of the head and body measurements. The total number of variables was 22, including 5 fingerprint patterns, 4 indices of the palm main lines, and 1 thenar and 1 hypothenar ridge pattern in the palm of each hand. The scoring system for each variable was mentioned at the beginning of this chapter: for each sex one run contained the Atayal, Bunun, Paiwan, Rukai, and Ami tribes, and the other the Atayal, Saisiat, Tsou, Puyuma, and Ami tribes. For the purpose of comparing these classifications with those based on the body and head measurements, the same tribes were contained in each run. The results are converted into percentages in Tables 47 and 48. The maximum percentage of people classified into their own tribe is 50, for the women of Tsou; the least percentage is 25, for the men of Rukai. It is quite obvious that the percentages of correct classification are smaller and those of misclassification greater for dermatoglyphics than for body and head measurements. Thus discrimination based on the dermatoglyphic data is not as efficient.

Factors contributing to the less discriminant power of dermatoglyphics compared to that of anthropometric measurements in our study may be (1) a relatively small number of genetic units responsible for the variations, so that large sampling errors are compounded, and (2) less adaptivity. The effects of natural selection may be greater on physical characters than on dermatoglyphics. Natural selec-

Table 47. Classification matrix based on the discriminant-function analysis of 22 dermatoglyphic variables on the fingers and palms of people of the Taiwan aboriginal tribes.[a]

Group	Atayal	Saisiat	Tsou	Puyuma	Ami	Total number of individuals
			Male			
Atayal	38	18	18	14	11	92
Saisiat	19	33	23	15	9	99
Tsou	13	26	48	6	6	99
Puyuma	14	15	18	32	20	65
Ami	17	11	17	13	42	142
			Female			
Atayal	32	14	26	14	13	145
Saisiat	12	43	17	17	11	97
Tsou	12	12	50	14	12	92
Puyuma	11	11	10	49	19	90
Ami	22	10	7	17	43	81

[a] The values in the 5 × 5 matrix table are the percentages of people of each tribe classified to their own and the other tribes. The values on the main diagonal represent percentages of right classification and those off the diagonal misclassifications. For example, in the males of the Atayal tribe, 38 percent of the people was classified to their own tribe, and 18, 18, 14, and 11 percent were classified to Saisiat, Tsou, Puyuma, and Ami tribes, respectively. The percentages should add up to 100. (Similar explanations apply to Table 48.)

Table 48. Classification matrix of five aboriginal tribes based on discriminant analysis of the dermatoglyphics of fingers and palms of people, showing the percentages of people classified into its own and other tribes.

Group	Atayal	Bunun	Paiwan	Rukai	Ami	Total number of individuals
			Male			
Atayal	46	16	17	9	12	92
Bunun	22	40	18	10	9	141
Paiwan	13	15	39	14	19	150
Rukai	16	20	19	25	22	102
Ami	12	13	18	13	43	142
			Female			
Atayal	38	17	20	12	13	145
Bunun	25	41	13	9	11	150
Paiwan	11	18	45	14	13	150
Rukai	15	20	15	29	21	80
Ami	20	16	14	6	44	81

tion has magnified the differences between tribes more in physical measurements than in dermatoglyphics and caused some variations in patterns.

MAHALANOBIS' D^2

Mahalanobis' D^2 was computed between each two tribes, using the scores for the 22 variables as above for the discriminant analysis and the same function as in analysis of the anthropometric traits. The results are given in Table 49 for men and Table 50 for women. There are slight discrepancies between women and men in some individual D^2's, but in general the D^2's for women are greater than for men, and the relative order of magnitude of the D^2's for the tribes agrees fairly well between the sexes. An average D^2 was computed for each tribe by taking the arithmetic mean of the seven D^2's of one tribe with each of the other tribes. These averages are given in the last lines of the tables.

For the Atayal tribe the D^2's vary greatly. There is a great difference between women and men in the D^2 to the Puyuma tribe. When both sexes are taken into consideration the relative distance to the Ami appears to be greater than those to the Saisiat, Bunun, and Paiwan. As far as the D^2's for the Saisiat are concerned, all values for women are higher than those for men. The D^2's to the Puyuma and

Table 49. Mahalanobis' D^2 between the Taiwan aboriginal tribes, based on dermatoglyphics of fingers and palms of men.

Tribe	Atayal	Saisiat	Bunun	Tsou	Paiwan	Rukai	Puyuma	Ami
Atayal		0.788	0.734	1.024[a]	1.054[b]	1.122[a]	0.945	1.085
Saisiat			0.728	0.727	0.727[a]	0.856	1.284[a]	1.256[a]
Bunun				0.633	0.947[b]	1.080[b]	1.886[b]	1.408[b]
Tsou					0.901[a]	1.426[b]	2.092[a]	1.590[b]
Paiwan						0.553	1.252[b]	0.778[b]
Rukai							1.288[a]	0.561
Puyuma								0.859
Average	0.963	0.940	1.012	1.244	0.885	0.999	1.350	1.103

[a] $P < 0.05$.   [b] $P < 0.01$.

Table 50. Mahalanobis' D^2 between the Taiwan aboriginal tribes, based on dermatoglyphics of fingers and palms of women.

Tribe	Atayal	Saisiat	Bunun	Tsou	Paiwan	Rukai	Puyuma	Ami
Atayal		0.916[b]	0.719[b]	0.912[a]	0.891[b]	1.047[a]	1.804[b]	1.481[b]
Saisiat			0.976[b]	1.285[b]	1.445[b]	1.485[b]	1.787[b]	1.864[b]
Bunun				0.678	1.135[b]	1.332[b]	1.836[b]	2.055[b]
Tsou					1.002[b]	1.944[b]	2.196[b]	1.897[b]
Paiwan						0.927[b]	1.168[b]	1.551[b]
Rukai							0.938	1.380[a]
Puyuma								1.528[b]
Average	1.107	1.350	1.282	1.345	1.165	1.287	1.569	1.681

[a] $P < 0.05$.   [b] $P < 0.01$.

Ami tribes are the highest for both sexes. It may be said that the D^2's to the Atayal, Bunun, and possibly Tsou and Paiwan tribes, are smaller than the distances to the other tribes. For the Bunun the D^2's to the Tsou are very small and those to the Puyuma and Ami are about three times greater in both sexes; the other D^2 values are intermediate. The D^2's of Tsou to the other tribes are rather similar to those for the Bunun in both order and magnitude. The D^2 for the Paiwan and Rukai in men is the smallest among all the D^2's, and the value in women is also quite low. This again indicates the close relationship between Paiwan and Rukai as revealed by the study of body and head measurements. The D^2 between Rukai and Ami in women is much smaller than that in men. The D^2's of the Ami and Puyuma to each of the other tribes are high in general, as shown by the average D^2's, indicating again that they are not closely related to the other tribes.

SUMMARY AND CONCLUSION

We have shown that significant differences exist in both fingerprint patterns and main line indices among the aboriginal tribes of Taiwan. If we assume that the ancestors of the present tribes derived from a common source, the significant differences among the tribes indicate the effects of isolation and of evolutionary change. They could also suggest that some tribes may have derived from different origins, especially the Ami tribe, which contributed the largest chi squares to the total for the eight tribes.

It should be noted that the total chi squares for the whorls and ulnar loops of the eight tribes are greater than those for the arches and radial loops, and between the whorls and ulnar loops the chi square for the whorls is greater. This indicates their relative importance in studying relationships between the tribes. Thus the use of the whorl/ulnar ratios to express tribal relationships may be considered a simple and reasonable method.

In the palm main lines the A line has larger chi squares than any of the other lines. This is not given in Table 42, but can be seen in Figs. 34 and 35. As for the fingerprint patterns, this is an indication of the effect of isolation and evolutionary change. Its relatively greater importance in sexual and bimanual differences is also indicated from the chi-square values in Table 43. I have no physiologic explanation for the relatively larger variation of the A line among the tribes and between sexes and hands except that in view of the terminal regions of the main lines in the palm, the A line apparently has more freedom to travel and hence more variation in its terminal regions, and is less transversary than the other lines. But the effect of evolution in relation to these conditions is difficult to visualize.

One of the most interesting aspects of the study of dermatoglyphics relates to bimanual and sex differences. There are greater frequencies of radial loops and whorls and less of arches in the right hand than in the left. This tendency exists in men and to some extent in women for practically all races (Kimura, 1962). In the palms the course of the main lines of right hands tends to be more transversal than that of left hands. This tendency was also more pronounced in men than in women. Bimanual differences are greater than sexual differences. In the same hand men tend to have more whorls and more transversal alignment of the main lines than women, especially in the right hands. These differences are clearly shown in our data. As most people are right-handed, the greater frequencies of whorl and radial loops and more transversal of the main line alignments in right hands are a consequence of evolution. Improvement of performance as a result of this is difficult to visualize, but there may be some, however minute. Further evidence pointing to a relationship between function and ridge patterns is that the thumb and index finger, which perform more varied tasks than the other fingers, show more variability in ridge patterns. In regard to the main lines, in some primates, such as the gibbon and the great apes, the main line alignments are more longitudinal than transversary. Wilder (1916) and Kanaseki et al. (1939) (quoted by Cummins and Midlo, 1961, p. 174) observed some human palms with remarkable similarity to the longitudinal configuration in apes. They regard the longitudinal alignment as an atavism. The bimanual differences suggest that as an evolutionary consequence the shift of the main lines is perhaps from longitudinal to oblique and transversary. This shift possibly indicates that the performance of the palm in early man may have been different from that in modern man.

Should the sexual and bimanual differences here mentioned be the results of evolution caused by functional difference, the claim of Cummins and Midlo (1961) that a race having a higher frequency of whorls would be less civilized is unjustified. They based their argument on the greater frequency of whorls in some species of primates. But since the preservation of certain primitive traits in the human species depends upon use, they are not always good criteria for determining primitivity.

As discriminant function is closely related to Mahalanobis' D^2 in mathematic theory, characters of less discriminant value reduce the reliability of the distance analysis. Nevertheless, the outcome of the distance analysis by means of Mahalanobis' D^2, except for a minor discrepancy, agrees quite well in essence with that based on the anthropometric characters; for example, the Ami tribe appears isolated from the other tribes, but the Rukai and Paiwan seem closely related. Such multivariate analyses of two dif-

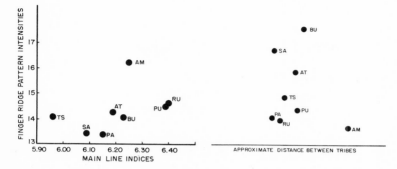

Fig. 37. (*Left*) Point diagram showing the relative "distance" between the tribes, based on the D^2's and plotted by approximation. (*Right*) Point diagram showing the relative separation between the tribes, based on the average fingerprint pattern intensities and the average main line indices of the people of the different tribes.

ferent sets of characters, which have different genetic bases and adaptive significance, can certainly not be expected to produce strict agreement in the results.

To illustrate the relationship between the tribes based on the D^2 analysis of the dermatoglyphics, a point diagram (Fig. 37, left) was constructed like the one for head and body measurements. In order to project a multidimensional plot onto two-dimensional space it was necessary to approximate the distances. In addition, a rectilinear diagram (Fig. 37, right) was based on the average intensities of fingerprint patterns and main line indices of the people in the different tribes. This plot is two-dimensional, one dimension representing the digits and the other the palms. A good deal of discriminatory power is thereby lost as compared with the D^2's based on a larger number of variables, but the gain in simplicity and precision of construction merits the use of this approach.

♀♂

CHAPTER 10. INTELLIGENCE

Intelligence is a most difficult trait to define and analyze; a vast controversial literature has grown up around intelligence tests and their meaning. General opinion has come to assume that, within certain limits, I.Q. and other behavioral tests do reliably measure certain mental abilities.

There is experimental and observational evidence that intelligence is determined partly by heredity. For instance, the classical work of Tryon (1940) showed that there are genetic differences in maze-learning ability in rats. In other animals there are large variations in mental capacity within species: different breeds of dogs vary greatly in alertness and performance; different mouse strains differ in learning capacity and other behavioral characteristics. In human beings the difference in I.Q. between relatives increases with decreasing genetic relationship (Hermann and Hogben, 1933). The work of Newman et al. (see Dobzhansky, 1962) showed higher correlation coefficients of intelligence between identical than between fraternal twins. The recent report of Shields (1962) showed less difference in the average intelligence scores between monozygotic twins than between dizygotic twins, and between monozygotic twins brought up together than between those brought up apart. Children's levels of mentality showed variation corresponding to the levels of mental ability required by their fathers' occupations (Scheinfeld, 1950).

Many test results have shown differences in I.Q. between

different races or ethnic groups, and differences on a racial or populational level have caused controversy and aroused prejudices. It is always legitimate to ask whether there is a genetic component in racial differences, since environmental components are difficult to evaluate. Even when people of different races inhabit the same area there is a question whether social conditions vary, and whether test results may be influenced by the social environment.

But that heredity plays a certain role in individual intelligence there remains little doubt. According to Dobzhansky (1962), "The academic lag goes far to explain why so many social scientists are repelled by the idea that intelligence, abilities, or attitudes may be conditioned by heredity. To suppose that a sex cell transports a particle called 'intelligence' which will make its possessor smart and wise no matter what happens to him is, indeed, ridiculous. But it is evident that the people we meet are not all alike in intelligence, abilities, and attitudes, and it is not unreasonable to suppose that these *differences* are caused partly by the natures of these people and partly by their environments." Attempts have been made to build a theory of the genetics of human intelligence. Hurst (1932, 1934) postulated that mediocrity was a dominant character dependent on a single major gene, whereas high or low intelligence was recessive. Later Pickford (1949) proposed a simple multifactor hypothesis based on the distribution of Stanford-Benet I.Q. scores in a general population. Neither one of these theories has been considered satisfactory (Fuller and Thompson, 1960). Perhaps intelligence behaves like some complex biologic trait such as, say, body growth, and is determined by many genes, with pleiotropic effects and interactions.

The main purpose of the present study of intelligence in the Taiwan aborigines was not to find the absolute level of mentality, in order to compare it with other races, but rather to find the relative differences between the tribes themselves, and to see how they might have resulted from long-term social and geographic isolation, and from differences in degree of civilization. The relative variability of such mental traits compared to that of physical and physiologic traits in these ecologic populations of man is our main interest here.

SUBTESTS

The Elementary School Intelligence Test, type II, forms A and B, was given to school children in the aboriginal districts. This is a pictorial test developed by the Ministry of Education of the Republic of China in 1961 as a modification of the Army General Classification Test. It has been found to be highly efficient and reliable in preliminary trials followed by comprehensive tests of more than 10,000 Chinese pupils in the elementary schools of Taiwan (Ku et al., 1961).

The test consists of three subtests: *figure discrimination, figure substitution,* and *block counting.* Each subtest has two parallel forms, A and B, in each of which the materials of the subtest are arranged slightly differently. The school children were given both forms of all three subtests, which, in separate booklets, were handed to them in succession. Each individual's total score for the three subtests represents his intelligence score.

Figure discrimination is a test of ability to select from among many symbols pairs that are identical (Fig. 38, top). The subject writes the letter O on the dotted line if the symbols match, and X if they do not. The test contains 80 such pairs. Testing time: 6 minutes.

Figure substitution is mainly a test of perception speed. The subject is shown two rows of symbols arranged in squares one above the other, at the top of the first page of the test.

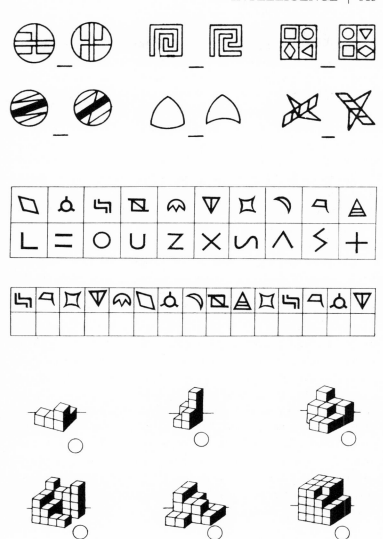

Fig. 38. (*Top*) Some of the figures used for the discrimination test; (*center*) some of the figures used for the substitution test (Ku et al., 1961); (*bottom*) some of the block piles used for the block-counting test.

Below this are two more rows, also marked off into squares, the upper containing the same symbols as in the top row above, but in different order, and the lower row is blank. The subject is asked to fill in the blank squares in the lower row with symbols that will produce pairs like those in the reference. A total of 240 blank squares are to be filled in (Fig. 38, center). Testing time: 8 minutes.

Block counting consists of 50 three-dimensional prints of block piles differently arranged, with the total number of blocks in each pile ranging from 4 to 64 (Fig. 38, bottom). The size and shape of the blocks are the same in all piles. The subjects are asked to figure the number of blocks in each pile and write the totals in the circle below. Testing time: 10 minutes.

The tests were given in uniform procedures throughout the schools by Mr. T. M. Lin, assisted by a school teacher. Mr. Lin explained the tests by means of an illustration with large print tacked to the blackboard. A few minutes were allowed for questions, and a 2-minute practice period was given in which items similar to those in the test were printed on the back of the test booklet. The instructor showed the correct answers, and another short question period was allowed. This completed the instruction. The subjects then started the tests at the same time and stopped at the same time. When form A had been completed there was a short rest period, after which form B was given. Testing times and procedures were the same as those used by Ku et al. in testing Chinese pupils. Scoring was 0.2 points for each correct answer in the figure substitution test, and one point for each correct answer in the other two.

ANALYSIS OF RESULTS

Mean ages. The average ages and total number of children receiving the different subtests in each tribe, and the number of schools in which they were distributed, are shown in Table 51. The average age ranged from 10.3 in the Puyuma to 11.1 in the Yami tribe for girls; and from 10.4 in the Ami to 11.1 in the Bunun tribe for boys. The differences between tribes appear to be rather small, but analysis of variance showed them to be significant in both sexes. The number of schools in which the tests were given ranges from two in the Saisiat to five in the Paiwan and Ami tribes. In most of the schools the children were of one particular tribe only, but some had a few Chinese. All these schools were established primarily for the aborigines except one in the Chinese section closely neighboring the Tsou tribe, which is primarily for the Chinese.

Test for normality. Before proceeding with the statistical analysis of the different test scores it was necessary to examine the distribution of the test scores for normality. For this the scores of Paiwan girls were used, because this tribe had the largest sample size, and those of the other tribes were

Table 51. Mean ages with standard errors of the children of the aboriginal tribes receiving intelligence tests.

Tribe	No. of schools distributed	Female		Male	
		No.	Mean ± s.e.	No.	Mean ± s.e.
Atayal	4	49	10.9 ± 0.16	58	11.0 ± 0.14
Saisiat	2	49	10.5 ± 0.14	46	10.9 ± 0.14
Bunun	2	82	10.8 ± 0.12	92	11.1 ± 0.12
Tsou	4	53	10.9 ± 0.03	81	10.5 ± 0.16
Paiwan	5	201	10.8 ± 0.10	220	10.9 ± 0.10
Rukai	4	58	10.3 ± 0.18	67	10.4 ± 0.13
Puyuma	4	105	10.1 ± 0.10	151	10.4 ± 0.10
Ami	5	109	10.2 ± 0.12	145	10.4 ± 0.11
Yami	2	62	11.1 ± 0.17	76	10.8 ± 0.14
Chinese	14	451	10.6 ± 0.06	445	10.6 ± 0.06

too small to give reasonable results. It is assumed that the distribution pattern of the scores in the different tribes is alike. A diagram of frequency distribution of the scores for each of the three subtests, form A, was constructed for Paiwan girls (Fig. 39). The distribution for the discrimina-

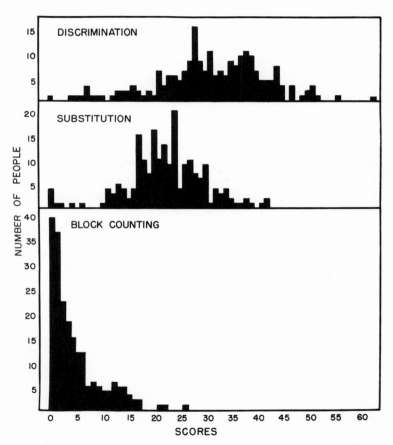

Fig. 39. Distributions of the discrimination, substitution, and block-counting subtest scores in all ages of the Paiwan females.

tion and substitution test scores approximates a bell-shaped curve, but for the block counting it is skewed.

The normality test for the discrimination and substitution subtests scores of the Paiwan girls was performed according to the method of Pearson and Hartley (1956), i.e., testing for kurtosis and skewness by analysis of the third and fourth moments. The moments of the observations are defined as

$$\bar{x} = \sum_{i=1}^{n} \frac{x_i}{n}, \qquad m_r = \sum_{i=1}^{n} \frac{(x_i - \bar{x})^r}{n} \ (r > 2),$$

where x_i is the score of ith individual in a sample of n individuals and m_r is the rth moment, and the well-known moment ratios are

$$\sqrt{b_1} = m_3/m_2^{3/2}, \qquad b_2 = m_4/m_2^2.$$

Departure of $\sqrt{b_1}$ from the normal value of zero is an indication of skewness in the frequency function of the sampled population, and departure of b_2 from the "normal" value of 3 is an indication of kurtosis. The symmetrical platykurtic distribution with $b_2 < 3$ is characterized by a flatter top and more abrupt terminals than the normal curve, while the symmetrical leptokurtic distribution with $b_2 > 3$ has a sharper peak at the mean and more extended tails. The results of the computation for the discrimination scores are

$$b_1 = -0.3276 \qquad \text{and} \qquad b_2 = 3.3332.$$

With a sample size of 201, b_1 is not significant; neither is b_2, but b_2 is greater than 3, indicating that the distribution is slightly leptokurtic. Therefore the sample is considered normally distributed. The test of normality for the substitutions subtest scores gives

$$b_1 = 0.0534, \qquad b_2 = 3.6386.$$

For $n = 197$, $\sqrt{b_1}$ is not significant but b_2 is slightly over the 5 percent level (3.57) but under the 1 percent level (3.98) of probability. These values show that the distribution of the substitution scores of Paiwan girls is very symmetrical but leptokurtic, which may be the result of some unequal age distribution in this subtest.

The extremely low mean scores for the block-counting tests indicate that the frequency distributions are likely to be very skewed. Indeed this tendency is shown clearly in the diagrams (Fig. 39) but the trend toward score increase with age also exists. The general distribution pattern appears to be of a Poisson type but it cannot be fitted by a Poisson distribution; mainly the observed frequencies in the 0 and 1 score categories are in excess of the expected frequency. In attempting to fit the Poisson distribution different square root transformations were made of the scores but were not successful, and it was decided to analyze the data by nonparametric methods.

Correlation analysis. Correlation of the results obtained from the two parallel forms is an estimate of reliability. Thus a product moment correlation analysis was made between form A and form B of each subtest and the combined scores of the three subtests by lumping all the subjects of the nine tribes together. The correlation coefficients obtained are given in Table 52 together with the correlation coefficients obtained by Ku et al. for the same tests administered to Chinese school children. For 1703 degrees of freedom, the correlation coefficients for the present study are all highly significant. Apparently the coefficients for the substitution and block-counting subtests and the three subtests together are very comparable in magnitude, and are greater than those for the discrimination subtest.

Regression of age on scores. In checking the effect of age and adjusting the mean scores of the tribes for age difference,

Table 52. Correlation coefficients between form A and form B of each subtest and the three subtests combined, given to the aborigines, together with the correlation coefficients obtained by Ku et al. for the same tests given to Chinese.

Tests	The present study	Ku et al. (1961)
Discrimination	0.509	0.581
Substitution	.765	.737
Block counting	.767	.774
Total scores	.752	.784

analysis of covariance was made on the scores of the discrimination and substitution subtests, following the procedures outlined in Snedecor (1957). The slight leptokurtosis in the distribution of the substitution scores is assumed to cause no serious effect in this and the following analyses. All the analyses were made with one run for each sex and each subtest, a total of eight runs. In each analysis the individual regression coefficients for each tribe of the score on age, together with the sum of the squares of deviation from each individual regression, are computed first. The regression coefficients range from 0.141 in the substitution subtest for the Atayal girls, to 4.648 in the discrimination subtest for Ami girls. The mean square of the deviation from individual regressions is used to test the differences among individual regressions and tribal means. If the differences among individual regressions are not significant, then there must be a common regression for all the tribes, the means can be adjusted for age differences, and tests for differences among adjusted and unadjusted tribal means can therefore be made. But if the differences among individual regressions are significant, testing for differences among tribal means becomes worthless. This method was considered for use in testing tribal differences.

Of the eight analyses, only three showed no significant differences in individual regression among the tribes. Therefore the results indicate an interaction between age and the test scores, and suggest that the covariance analysis is not a proper test for the tribal differences.

Sexual differences. The mean and standard error for each subtest were computed on a school grade basis for each sex of each tribe (Tables 53–55). Third-grade students were omitted from the analysis because no test was given to pupils of this grade in the Atayal, Saisiat, and Yami tribes, and the number of third graders in the other tribes was relatively small. The average of all grades for each tribe and that of all tribes for each grade are also given in the same tables. They show that in general boys were superior to girls in the discrimination and block-counting subtests, but not in the substitution subtests. In the mean scores of the substitution subtests for each grade, sexual differences are rather small in comparison with differences among tribes. For the block-counting subtest, sexual differences are relatively greater than in the other subtests, and are consistent through the tribes. The variation of sexual differences among the tribes suggests the effects of socioeconomic environment on a particular test performance, and the relative magnitude of variation reflects the relative variation of such environment among the tribes.

Tribal differences. The mean scores in each subtest vary among the tribes (Tables 53–55). In the discrimination subtest, the Bunun has the lowest mean score and the Puyuma tribe the highest for both sexes. In the substitution subtest, the Yami has the lowest and the Ami tribe the highest mean score for both sexes. In the same subtest, the Tsou has the same mean score as that of Ami for males. In the block-counting subtest, the Yami tribe again ranks lowest in mean scores, and the Rukai highest. In the same test, Saisiat

Table 53. Means and standard errors of the figure discrimination subtest scores of the aboriginal tribe pupils in the different school grades.

| | | Test Type A | | | | | | Test Type B | | | | | | |
| | | Grade 4 | | Grade 5 | | Grade 6 | | Grade 4 | | Grade 5 | | Grade 6 | | |
Tribe	Sex	N	Mean ± s.e.	N	Mean ± s.e.	N	Mean ± s.e.	N	Mean ± s.e.	N	Mean ± s.e.	N	Mean ± s.e.	Ave.
Atayal	♀	17	29.6 2.11	15	31.9 3.32	17	36.2 2.90	17	37.5 2.50	15	40.6 1.68	17	38.1 1.90	35.7
	♂	15	30.7 2.23	26	32.8 2.57	17	41.2 2.54	15	37.2 2.09	26	39.9 1.75	17	41.8 1.58	37.3
Saisiat	♀	24	34.1 1.80	18	30.5 1.74	7	32.8 2.45	24	36.5 1.11	18	38.9 2.16	7	33.7 3.11	34.4
	♂	12	34.8 2.20	22	36.4 2.29	12	42.8 2.72	12	34.2 2.06	22	38.0 1.74	12	44.0 2.96	38.4
Bunun	♀	26	25.7 2.38	36	30.2 1.70	20	35.4 2.65	26	35.3 1.38	36	35.2 1.44	20	40.0 1.63	33.6
	♂	26	33.7 2.18	31	33.6 1.74	36	35.9 1.67	26	37.7 1.28	31	37.9 1.35	36	38.8 1.60	36.3
Tsou	♀	12	33.8 1.90	19	28.9 2.38	10	43.9 1.75	12	42.2 2.40	19	38.0 2.04	10	45.9 1.97	38.8
	♂	19	32.2 2.67	23	34.6 1.73	15	39.7 2.27	19	36.2 3.23	23	39.3 1.53	15	41.0 2.17	37.2
Paiwan	♀	49	29.1 1.25	72	33.0 0.96	46	37.2 1.42	49	35.7 0.97	72	37.2 0.87	46	41.4 1.33	35.6
	♂	57	33.4 1.24	72	34.8 1.02	52	37.8 1.13	57	35.6 1.00	72	37.8 0.96	52	43.4 1.20	37.1
Rukai	♀	12	24.9 2.68	11	38.1 4.32	11	40.0 1.91	12	34.8 2.40	11	40.5 2.72	11	39.4 2.64	36.3
	♂	22	35.3 1.68	10	41.1 1.95	9	42.2 2.02	22	35.9 1.60	10	42.0 2.44	9	41.6 1.84	39.7
Puyuma	♀	29	33.3 1.57	26	38.0 1.74	13	36.8 2.17	29	39.1 1.43	26	40.7 1.79	13	41.1 2.25	38.2
	♂	44	34.8 1.29	34	38.0 1.50	29	39.7 1.73	44	38.1 1.27	34	42.3 1.42	29	46.4 1.78	39.9
Ami	♀	23	32.3 1.78	25	35.2 1.76	20	35.6 2.00	23	37.6 1.66	25	41.8 1.59	20	43.0 1.53	37.6
	♂	40	32.0 1.39	21	39.8 1.81	36	41.4 1.72	40	38.9 1.31	21	42.0 2.06	36	42.5 1.44	39.4
Yami	♀	—	—	28	34.9 1.49	—	—	—	—	28	36.2 1.51	—	—	35.6
	♂	—	—	31	38.1 1.20	—	—	—	—	31	39.4 1.31	—	—	38.8
Average	♀		30.4		34.7		37.2		37.3		38.6		40.3	
	♂		33.4		36.7		40.1		36.7		39.7		42.4	
Chinese	♀	134	33.4 0.83	130	36.5 0.74	124	40.0 0.76	134	36.9 0.60	130	40.4 0.70	124	43.2 0.71	38.4
	♂	125	35.6 0.69	124	37.9 0.74	122	41.4 0.86	125	37.6 0.76	124	39.5 0.77	122	44.9 0.75	39.5

Table 54. Means and standard errors of the figure substitution subtest scores of the aboriginal tribe pupils in the different school grades.

Tribe	Sex	Test Type A									Test Type B									Ave.
		Grade 4		Grade 5		Grade 6			Grade 4		Grade 5		Grade 6							
		N	Mean ± s.e.	N	Mean ± s.e.	N	Mean ± s.e.		N	Mean ± s.e.	N	Mean ± s.e.	N	Mean ± s.e.						
Atayal	♀	17	20.7 1.90	15	30.9 1.83	17	28.6 3.49		17	24.9 1.81	15	33.2 2.21	17	27.6 3.23						27.7
	♂	15	22.7 1.64	26	25.7 2.52	17	33.0 1.84		15	24.7 1.71	26	25.8 1.44	17	30.0 1.63						27.0
Saisiat	♀	24	23.4 1.45	18	18.7 1.19	7	22.1 1.45		24	24.2 1.32	18	22.7 1.53	7	27.6 1.80						23.1
	♂	12	19.8 2.01	22	19.9 1.45	12	25.3 1.49		12	20.8 2.35	22	24.0 1.38	12	32.4 1.90						23.7
Bunun	♀	26	17.2 1.73	36	15.5 1.64	20	23.5 2.11		26	21.7 1.63	36	17.8 1.73	20	27.6 1.36						20.6
	♂	26	20.5 1.85	31	18.1 1.61	36	25.7 1.20		26	25.0 1.76	31	20.0 1.82	36	29.0 1.50						23.1
Tsou	♀	12	21.2 1.98	19	29.4 1.78	10	32.3 1.84		12	24.6 1.87	19	26.7 1.53	10	36.5 2.12						28.5
	♂	19	19.2 2.02	23	26.8 1.52	15	30.5 1.56		19	23.4 2.42	23	30.8 1.54	15	33.5 1.67						27.4
Paiwan	♀	48	22.0 0.81	72	22.8 0.84	46	27.1 0.96		48	25.1 1.00	72	27.2 1.26	46	30.2 0.95						25.7
	♂	57	21.8 0.73	72	24.2 0.75	52	27.4 0.95		57	24.3 0.98	72	27.9 0.98	52	30.7 1.26						26.1
Rukai	♀	12	19.1 2.10	11	21.7 3.93	11	29.1 1.89		12	24.7 2.53	11	30.5 3.63	11	35.5 2.18						26.8
	♂	22	16.7 1.66	10	28.6 2.26	9	26.8 3.02		22	23.0 2.17	10	32.6 2.22	9	30.8 3.10						26.4
Puyuma	♀	29	21.4 1.05	26	25.6 1.35	13	27.2 1.34		29	26.8 1.44	26	28.0 1.46	13	31.2 1.65						26.7
	♂	44	20.8 0.77	34	25.1 1.43	29	25.1 1.00		44	23.9 0.90	34	25.7 1.45	29	28.7 1.14						24.9
Ami	♀	23	21.2 1.15	25	27.0 1.05	20	30.4 1.40		23	26.0 1.50	25	34.7 1.43	20	35.4 1.59						29.1
	♂	40	22.4 0.84	21	25.6 1.49	36	28.5 0.91		40	24.5 1.06	21	30.8 1.70	36	32.7 1.09						27.4
Yami	♀	—	—	38	20.9 1.23	—	—		—	—	38	23.5 1.40	—	—						22.2
	♂	—	—	31	20.8 1.07	—	—		—	—	31	26.2 1.11	—	—						23.5
Average	♀		20.8		23.9		27.5			24.8		26.9		31.5						
	♂		20.5		24.5		27.8			23.7		26.5		31.0						
Chinese	♀	134	22.0 0.66	130	25.8 0.69	124	27.0 0.69		134	25.4 0.61	130	29.5 0.72	124	31.4 0.69						26.9
	♂	125	20.7 0.55	124	24.2 0.73	122	27.5 0.63		125	23.9 0.75	124	27.7 0.73	122	31.5 0.63						25.9

Table 55. Means of the block-counting test scores of the aboriginal tribe pupils in the different school grades.

Tribe	Sex	Test Type A						Test Type B						Average
		Grade 4		Grade 5		Grade 6		Grade 4		Grade 5		Grade 6		
		N	Mean	N	Mean	N	Mean	N	Mean	N	Mean	N	Mean	
Atayal	♀	17	3.41	15	4.20	16	4.31	17	5.06	15	5.60	17	5.94	4.75
	♂	15	5.46	26	6.00	17	11.35	15	5.27	26	8.85	17	8.70	7.60
Saisiat	♀	24	5.25	18	4.61	7	5.85	24	8.38	18	7.89	7	8.43	6.74
	♂	12	7.83	22	7.86	12	11.33	12	9.33	22	7.54	12	13.92	9.64
Bunun	♀	26	2.11	36	4.05	20	7.30	26	3.00	36	4.89	20	9.60	5.16
	♂	26	6.11	30	5.93	36	8.05	26	5.92	30	6.53	36	11.25	7.30
Tsou	♀	12	3.50	19	3.21	10	4.80	12	4.42	19	5.68	10	7.00	4.77
	♂	19	6.26	24	7.37	15	6.00	19	7.16	24	7.08	15	7.27	6.86
Paiwan	♀	49	3.83	72	5.08	46	5.28	49	4.69	72	7.65	46	7.85	5.73
	♂	57	6.31	72	7.06	52	10.51	58	8.22	78	8.04	52	13.10	8.88
Rukai	♀	12	3.50	10	9.20	12	5.50	12	8.33	11	6.36	11	12.00	7.48
	♂	22	2.27	11	7.27	8	12.00	22	10.09	10	13.30	9	12.89	9.64
Puyuma	♀	29	2.37	26	5.38	13	9.30	29	3.48	26	9.27	13	11.46	6.88
	♂	44	7.52	34	6.20	29	11.41	44	8.02	34	7.59	29	13.28	9.00
Ami	♀	23	4.00	25	5.44	20	5.75	23	4.87	25	6.88	20	6.55	5.58
	♂	48	4.12	21	7.09	36	8.38	40	7.02	21	9.00	36	11.47	7.84
Yami	♀	—	—	28	2.69	—	—	—	—	28	3.54	—	—	3.11
	♂	—	—	31	5.34	—	—	—	—	31	6.33	—	—	5.84
x^2	♀	192	13.84	249	9.91	144	5.92	192	12.21	250	16.91[a]	144	8.58	—
	♂	243	14.40[a]	271	5.11	205	5.69	236	9.38	276	9.45	206	15.02[a]	—
Chinese	♀	134	6.29	130	7.94	124	9.53	134	8.49	130	11.19	124	13.54	9.50
	♂	125	8.22	124	10.26	122	13.91	125	10.63	124	12.38	122	16.56	11.99

[a] $P < 0.05$.

199

has the same mean scores as the Rukai for males. The mean scores of the Chinese are higher than practically any mean scores of the tribes in the block-counting subtests, but not in the others.

A mean score for each subtest is computed for each tribe, taking the arithmetic mean of all grade averages for both sexes and for both form A and form B; the total score for each tribe is the sum of the means of the three subtests (Table 56). These means and total scores, with equal weight for grades, are better estimates than the means and total scores based on individual records of each tribe. The Yamis, having only the fifth grade, which is intermediate between the fourth and sixth grades, should compare properly with the other tribes as far as the weight of grade is concerned. The tribes are ranked according to these means and total scores. The rank order based on the means for the discrimination subtest coincides well with that based on the substitution subtest, but neither agrees so well with the rank based on the block-counting subtest. In general the Amis rank high and the Bununs and Yamis low.

Table 56. Mean scores of each subtest and total mean scores, ranked according to primitivity and the mean scores for the tribes.

Tribe	Civili- zation rank	Discrimination		Substitution		Block counting		Total	
		Mean	Rank	Mean	Rank	Mean	Rank	Score	Rank
Atayal	1	27.4	3	36.5	5	6.2	7	70.1	5
Saisiat	3	23.4	7	36.4	6	8.2	2	68.0	7
Bunun	1	21.9	9	35.0	7	6.2	6	63.1	9
Tsou	2	28.0	2	38.0	3	5.8	8	71.8	4
Paiwan	2	25.9	5	36.4	6	7.3	4	69.6	6
Rukai	1	26.6	4	38.0	3	8.6	1	73.2	2
Puyuma	4	25.8	6	39.1	1	7.9	3	72.8	3
Ami	3	28.3	1	38.5	2	6.7	5	73.5	1
Yami	1	22.9	8	37.2	4	4.5	9	64.6	8
Chinese	—	26.4	—	39.0	—	10.75	—	76.15	—

Table 57. Summary of results of analyses of variance for differences between tribes in discrimination and substitution subtests.

Subtest		Grade	Sex	Mean squares		Degrees of freedom			Percent variance between tribes
				Between tribes	Within tribe	Between tribes	Within tribe	F	
Discrimination	A	4	♀	260.0	83.6	7	184	3.10[b]	8.1
	A	4	♂	57.8	86.4	7	227	0.66	
	B	4	♀	98.1	56.4	7	184	1.73	
	B	4	♂	65.9	71.2	7	227	0.92	
	A	5	♀	234.1	87.6	8	241	2.67[b]	5.9
	A	5	♂	180.6	84.1	8	261	2.15[a]	3.5
	B	5	♀	140.5	67.0	8	241	2.10[a]	4.0
	B	5	♂	102.2	65.9	8	261	1.55	
	A	6	♀	114.4	90.7	7	136	1.26	
	A	6	♂	139.9	86.5	7	198	1.61	
	B	6	♀	126.9	65.4	7	136	1.94	
	B	6	♂	151.7	77.4	7	198	1.95	
Substitution	A	4	♀	86.8	45.1	7	184	1.92	
	A	4	♂	91.9	43.8	7	227	2.09[a]	3.8
	B	4	♀	58.3	54.6	7	184	1.06	
	B	4	♂	88.2	61.3	7	227	1.43	
	A	5	♀	535.9	57.9	8	241	9.26[c]	23.8
	A	5	♂	282.9	62.5	8	261	4.53[c]	10.9
	B	5	♀	2034.1	83.1	8	241	24.47[c]	47.0
	B	5	♂	350.3	63.6	8	261	5.50[c]	13.5
	A	6	♀	140.4	64.1	7	136	2.19[a]	6.6
	A	6	♂	147.1	42.8	7	198	3.44[b]	9.0
	B	6	♀	205.9	57.9	7	136	3.55[b]	13.1
	B	6	♂	73.7	60.8	7	198	1.21	

[a]$P < 0.05$.　　[b]$P < 0.01$.　　[c]$P < 0.001$.

A variance analysis for these tests on either a school grade or an age basis may be pertinent. As the grade is highly correlated with age, and as there is no logical basis on which to set up an age interval, it was decided to run the analysis of variance on each individual school grade. This eliminates the effects of both age and grade.

One analysis of variance was made for each subtest for each sex in grades four, five, and six—a total of 12 runs (Table 57). The table contains the mean squares between

and within tribes, their ratios as given under F, and the between-tribe variance percentages, that is (variance between
tribes)/(variance between tribes + variance within tribes).
(The computation procedures are like those given in Chapter 6.) It can be seen that only 4 of a total of 12 mean square
ratios are significant in the discrimination subtests, and 8 of
12 in the substitution subtests. These are for the fourth-
grade girls and for both boys and girls of the fifth grade in
form A; fifth-grade girls only in form B are significant for
discrimination. For the substitution subtests, the ratios are
for boys of the fourth grade in form A, for both sexes in the
fifth grade for forms A and B, and in the sixth grade for
both sexes for form A, and for girls only in form B. The percentage of between-tribe variance ranges from 0 to 25.05.

A chi-square test of heterogeneity was used for the
analysis of possible tribal differences in the block-counting
subtest. It was performed with each grade separately as in
the analysis of variance in the other subtests. It is not possible to use more than two classes of scores in each grade of
each sex, for the numbers would be too small in some classes
to apply the chi-square tests. Therefore each sex, grade,
and test are divided into only two classes, and in such
a way that there should be no large differences between
totals in the two different classes. The range of scores included for the first class was 0-2 to 0-7 in the different runs.
It is rather obvious that the analysis is lacking in efficiency,
but there seems to be no statistical method that would provide a more sensitive test.

A total of 12 chi-square tests was made, one for boys and
one for girls in each of the fourth, fifth, and sixth grades.
(For the reasons explained above, the third grade was
omitted from the analysis.) The results (Table 55) show
that the chi square is significant in only 3 of the 12 tests,
i.e., for fourth-grade males in the form A subtest; and fifth-

grade females and sixth-grade males in the form B subtest. None of the remaining chi squares for testing the differences between tribes is significant. When all the scores of the tribe pupils are pooled together and tested against the corresponding Chinese scores the chi squares are all significant.

Relationship of intelligence and civilization. Since it appears that the test scores may have some relationship to the degree of civilization (see Chapter 2), the tribes are so ranked. A correlation analysis is made between these rank scores and the mean and total test scores for the tribes. The following correlation coefficients are thus obtained:

Discrimination	0.216,
Substitution	0.555,
Block counting	0.450,
Total score	0.493.

For seven degrees of freedom, none of them is significant. But as these values are all positive and fairly high, they possibly indicate a trend toward higher scores among civilized tribes. It must be kept in mind that the scores given according to degree of civilization are arbitrary and mainly for computational convenience.

A two-dimensional diagram (Fig. 40) showing distribution of the tribes was constructed by plotting the standardized average tribe scores for the discrimination test on the X axis and that for the block-counting test on the Y axis. Assuming no genetic effect on the distribution, the space so defined can be considered as a civilization space, the civilization being measured linearly on the diagonal line drawn from the lower left to the upper right corner. Excepting the Rukai tribe, a general agreement appears between the degrees of civilization estimated by the test scores and by the anthropologists. As mentioned previously, a portion of the sample of the Rukai tribe came from a village on the east side of Taiwan, where living conditions are similar to those in the neighboring tribes of Puyuma and Ami, which are considered rather advanced. This circumstance may have contributed to their high scores. Incidentally, the civilization scale shown in the diagram can be stated by the familiar formula

$$Z = \sqrt{x^2 + y^2},$$

where Z = civilization, and x and y are the two different standardized test scores. We realize that the actual condition should not be so simply expressed.

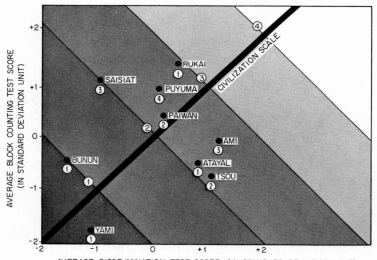

Fig. 40. Two-dimensional space diagram made by plotting the average block-counting test score against the average discrimination test score of each tribe. The number attached to each tribe is the civilization rank given according to the classification by Chen (see Chapter 3). The numbers in the circles are the civilization rank measured on a linear scale of the diagonal running from the lower left to the upper right corner of the diagram. Assuming no genetic differences in the test scores, the diagram shows that the estimate based on the two test scores is generally in agreement with the classification given by Chen, except for the Rukai tribe.

DISCUSSION AND CONCLUSION

The results of the correlation analyses of the three subtests applied to the aboriginal children indicate that figure substitution and block-counting subtests are more reliable than the figure discrimination, but the block-counting test was too difficult for the tribe children, causing skewed score distributions and zero scores for many subjects.

It is interesting to see that sexual differences vary among the subtests and that boys surpass girls in figure discrimination and block counting, but not in figure substitution. It is difficult to assess the relative importance of genetic and environmental influences on the differences; however, in the substitution subtest a large contrast appears between the Ami and Bunun tribes, with Bunun boys surpassing girls, and Ami girls surpassing the boys, both by wide margins, which seems to indicate that the social environment plays a much more important role than the genetic, or otherwise the difference in the Bununs would be in the same direction as in most of the other tribes.

When all the scores of the aborigines are pooled and compared with the scores of the Chinese children, the Chinese rank above the aborigines in all the tests, and the differences are practically all significant. Although these Chinese children are living in or near the aboriginal areas and attending the same schools, it seems unreasonable to rule out the possibility of environmental contribution to the total difference. This, however, as mentioned at the beginning of this chapter, is not our primary interest in the study, and we will discuss it no further here.

When the results here obtained are analyzed on a school-grade basis, no overwhelmingly significant differences appear between the tribes. Most of those that are significant at all occur in the fifth grade—the only grade containing data from the Yami tribe. As mentioned previously, the Yami is probably the least civilized of the Taiwan aboriginal tribes, a probability that appears to contribute most to the significance of the differences. The correlation analysis and the two-dimensional diagram also show some association between advancement of civilization and total test scores. The overall evidence seems to indicate that the slight differences in the within-grade tests can be attributed mainly to variation in social and economic environment.

As each local population has its own coadapted gene pool, and as the frequencies of genes relating to intelligence may vary between one gene pool and another, why do no definite differences appear among the tribes? Assuming that genes relating to intelligence are many, and have properties of additivity and pleiotropism, their average frequency in any two populations derived from one or related parent populations would not differ much at the beginning, even though there might be differences in individual gene frequencies. During isolation natural selection should be in favor of high intelligence in each of such primitive populations. Students of evolution have been deeply concerned with the effects of natural selection on intelligence. The basic problem arises from the fact, generally recognized, that environmental effects on the variation of intelligence are much greater than genetic effects, and that separation of these two components is difficult. This difficulty has caused argument as to how close the various test scores come to expressing the genetic influence on intelligence.

Many investigators take a pessimistic view (see Mayr, 1963), of the evolution of intelligence in modern and economically highly developed societies, because it appears that children from large families have lower I.Q.'s than those from smaller families. But this could be attributed to such social factors as individual children in large families receiving less care. Terman and Oden (1959) in connection with their study of exceptionally high I.Q.'s, pointed out that fashion rather than biology may sometimes dictate family size, e.g., in the United States in the 1920's it was fashionable for intellectuals to have small families, but then around 1950 the fashion changed and intellectuals began producing large families.

Based on his study and that of Bajema (1963) Reed (1965) pointed out that the negative correlation between the number of children in a family and the intelligence of the parents as usually claimed is arithmetically satisfactory but an artifact when evaluating the relationship of reproduction and intelligence, when the childless brothers and sisters of the parents used in the analysis were not included. Reed and Bajema showed that when the members of a generation who failed to reproduce were included, the parents with above-average intelligence had slightly more children than the below-average parents. Reed concluded that the failure to consider the intelligence of those in a generation who failed to reproduce provides a large statistical bias, or error, so that to assume that the mentally retarded produce more than enough to replace themselves, while the more intelligent fail to marry or have fewer children, is a misconception.

But whatever evolutionary changes may take place in highly advanced societies, the evolution of intelligence in less civilized populations, such as the Taiwan aboriginal tribes, can be quite a different thing. We may generally agree that in primitive societies the upper classes have the greater physical and mental abilities. For instance, chieftains or others at the higher social levels usually find it easier to obtain mates from a comparable class and possibly to raise larger families than people in the lower classes. This would of course be true for primitive groups practising polygamy; but in the Taiwan aboriginal tribes, although monogamy is practised as a rule, people socially or economically high up in the hierarchy tend to bring up more children, whether in the clan or the caste system. On the other hand, people of low intelligence may have difficulty in obtaining mates and raising families for socioeconomic reasons. Therefore, high intelligence should always be favored by natural selection in such societies, although there may be transient slowing down of evolutionary progress due to some social change or reform. Reed (1965) stressed the argument further and said, "Intelligence would seem to be an orthoselective trait of the first degree because it is unlikely that there is any environment where lower intelligence really would be advantageous to a person, and high intelligence disadvantageous." Some special faculties correlated with special physical and physiologic characteristics of a race or a group, such as athletics, musical talents, and a certain type of technics, may show differences on a racial level. As far as general intelligence is concerned, the basic promoting force, natural selection, has perhaps acted on each of the tribal populations in more or less the same way. Unless there had been a large difference in genetic endowment in the beginning, it is unlikely that there would be noticeable differences in intelligence among the tribes.

To summarize, among populations such as the Taiwan aboriginal tribes, which are assumed to have been randomly dispersed on the island at the beginning, from one or closely related parental populations, selection is in favor of high intelligence within each tribe. Thus genetic differences may show in individuals as in any other populations, but not at the tribal level. If there should be genetic differences among the tribes, they would be minute and overshadowed by the influences of the socioeconomic environment. For all practical purposes, the present mean scores of the intelligence tests for the tribes may be considered as reflecting to a large extent the relative civilization among the tribes.

♀♂

CHAPTER II. GENERAL CONCLUSION

I have recorded in this monograph a cross-sectional examination of various biologic characteristics in the Taiwan aboriginal tribes at a certain moment in their evolution. The customary technics of physical anthropology, psychology, and physiology were used in collecting the data, which cover both polymorphic and polygenic characters. Examinations, including anthroposcopic observations, anthropometric measurements, dermatoglyphics of hands, and tests for blood pressure and PTC taste thresholds, were given to about 1900 adults in eight tribes; intelligence tests were administered to 2600 school children in nine tribes. Biologic relationships among the tribes were established mainly on the basis of the anthropometric and dermatoglyphic data, which were analyzed by the discriminant function and Mahalanobis' D^2 technics. Only by using these statistical methods could the large numbers of variables for eight groups of people be efficiently utilized for classification.

A species or a population varies in time and space, and the study of human evolutionary processes involves both dead and living subjects, the latter providing the broader spectrum of characters for examination. In any attempt to trace the mechanisms of cause and effect involved in evolution, principles derived from the study of living man are needed. In our study of the existing Taiwan aborigines the cause of tribal differentiation is mainly isolation. What then are the effects of isolation on biologic variation? What are the possible genetic bases for the differences among the tribes?

The racial framework of Taiwan includes three ethnic groups: the now extinct Negritos, mentioned in Ching Dynasty documents (Chapter 3) and in aboriginal folklore (Chapter 5); the present aboriginal tribes; and the Chinese.

The widespread presence of Negritos in the South Pacific Islands supports the theory that they may have been the earliest inhabitants of Taiwan. Kroeber (1948) referred to them as an ancient and primitive people, who may have inhabited wide stretches of territory in Africa, Asia, and Oceania before its invasion (Coon, 1962) by Mongoloids and modern kinds of Caucasoids from the north. Our study provides also what may be biologic evidence of genetic infiltration from this vanished race, in the short stature and dark skin color of the western tribes, especially the Bununs and Paiwans, who also have the broad head and nose and short, wide face typical of the Negritos.

Next come the ancestors of the present aboriginal tribes, who migrated either from ancient tribes on the east coast of the mainland of China before 200 B.C. (Chapter 3) or from other places, such as the Malay Peninsula, the Philippines, and other islands in the South Pacific. There is no way in which our data can throw light on questions regarding these 2000-year-old origins.

The third and most recent group to inhabit Taiwan comprises the Chinese, who came first from the east coast of Fukian and Kwongtung Provinces, with migrations starting about 200 A.D. and continuing periodically to the present day.

Whatever the origin or origins of the present Taiwan aborigines may be, analysis of our data has revealed that tribal variations and relationships are closely associated with geographic and social isolation. The discriminant function analysis showed that in some cases as high as 80 percent of a sample could be classified into its own tribe. This fact demonstrates the efficiency of the mathematic function in handling large numbers of variables in classificatory problems on the one hand, and the effect of isolation in the tribal populations on the other. The analysis also revealed that biologic relationships among the tribes closely approximated geographic ones. For example, the Ami and Bunun, most isolated geographically, were found to be most widely separated biologically from the other tribes; whereas the Paiwan and Rukai, closely associated geographically, are also closest biologically.

The Ami tribe inhabits a long strip of land on the east coast of Taiwan, with the Pacific Ocean on the east and the high Central Mountain Range along the west, so that the territory is virtually sealed in with only the two narrow ends open. The geographic isolation of this tribe is actually comparable to or greater than the so-called "river or shore line territory" isolation (Falconer, 1961), which limits migration. There are sharp physical differences between the Ami and other surrounding tribes, such as the Bunun, Paiwan, and Rukai, the Ami people being comparatively tall, slender, and light colored; the latter are short, stocky, and dark. Many facial characteristics (Chapter 5) also show great dissimilarity, indicating large genetic differences.

The most interesting finding of the Mahalanobis' distance analysis concerns the relationship of the Bunun to the other tribes. The Bununs, a comparatively uncivilized and conservative tribe, are as isolated as the Amis, though for different reasons. They are located at about the geographic center of the tribal area, but at a much higher altitude. If the tribal distribution were visualized in three-dimensional space, the present analysis would locate the Bununs high above a plane on which all the others were scattered around and slightly up and down, a picture in close approximation to the topography of their distribution. Thus they are isolated both geographically and socially. This isolation correlates well with the wide biologic separation indicated with remarkable clarity by the D^2 analysis of the anthropometric data (Tables 27 and 28)—a separation that has not been previously reported.

These biologic findings regarding the separation of the Ami and closeness of the Paiwan and Rukai tribes confirm the findings of anthropologists, who used lingual and cultural criteria. There is disagreement, however, regarding the Saisiat and Atayal tribes, and the Ami and Puyuma; our study did not reveal the closeness claimed for them. Rather surprisingly, the Puyuma, an eastern tribe, and the Rukai, the majority of whose population inhabits the west, appear to be quite closely related, the biologic distance between them being less than between the Puyuma and its eastern partner, the Ami. The map shows that the geographic border contact between the Rukai and Puyuma is large considering the small area occupied by these tribes. It is very likely that as a consequence a relatively large mutual infiltration of genetic elements has taken place between these two tribes.

The results of the discriminant analysis showed large numbers of Atayals classified to other tribes, an indication of heterogeneity. The Atayals, widely spread through the northern mountain ranges, are in direct contact with the Saisiat, Bunun, and Ami, and also with the Chinese; furthermore they appear to be more aggressive than the other tribes (they fought the Japanese during the latter's occupation), and both these factors may have favored their intermingling with other groups.

During the long period of their isolation the various tribes have been undergoing evolutionary processes of all kinds, which have produced the present biologic distinctions found among them.

Our data show (Chapter 6) that the Bununs have the largest chest girth of any of the aborigines, indicating adaptation to high altitude and pointing to an effect of both normalizing and directional selection. This characteristic and others, such as short stature and compact body build, contrast with those of plains people like the Amis, who are tallest among the tribes and have the smallest chest girth. One may question whether the large chest girth of the Bununs is merely a physiologic adaptation, and I would not argue that no environmental component is involved. But it is my supposition that there is a genetic component in most anatomic characters, as shown in other species and in the recent study of twins by Osborne and DeGeorge (1959), and that natural selection often favors genetic fixation of traits of a type induced by a given environment (Dobzhansky, 1962). This phenomenon is sometimes referred to as "organic selection" or the "Baldwin effect." Simpson has devoted a rather full discussion to it, and the Russian workers Gause and Lukin, experimenting independently, arrived at the conclusion that natural selection is in favor of genes that act in the same direction as the environment (Waddington, 1957). Waddington has further postulated a theory of the genetics of development using the terms "canalizing selection" and "genetic assimilation" to describe such evolutional changes. But to demonstrate unequivocally that genetic changes have actually occurred concerning, say, the chest girth in the Bununs would require a control population, with offspring raised in a different environment for comparison with the individuals of the parental population. Such experiments are almost impossible in human groups.

We found that although the tribes' dermatoglyphic variations (Chapter 9) generally correspond with those of head and body measurements (Chapter 6) between tribes, the separations based on anthropometry were larger than those based on dermatoglyphics, thus indicating an effect of selection, that is, it would appear that natural selection has had greater influence on form and shape of the head and body than on dermatoglyphics of the hands, though we are aware that there may be an environmental portion of variance in the anthropometric measurements.

It is interesting to note evolutionary effects on variation in genetically less definable traits, such as blood pressure, which is genetically complex and strongly affected by natural and social environments. The blood pressure level among the Taiwan aborigines is in general low, with average systolic and diastolic pressures of 114 and 73 mm Hg respectively for both men and women of all the tribes, the lowest averages being 106 and 65 mm Hg for Ami women. The averages are adjusted to age 38 (the mean age of all adults examined). In civilized populations a reading of 120 mm Hg for systolic and 80 mm Hg for diastolic pressure is considered normal for 20 years of age. We also found that in some tribes, such as the Ami, the blood pressure appeared to show no increase with age, which is in line with some previously reported cases of hypotension (Chapter 8). This is likely to be a consequence of natural selection on a population of some Mongoloid genetic background, conditioned to some extent by environment.

The least genetically definable trait is intelligence, since it is highly affected by socioeconomic influences that are most difficult to assess, and the genetic component is generally small. Using a pictorial type of test, we found some significant differences among the tribes, but the variations do not correspond to tribal relationships indicated by the anthropometric and dermatoglyphic data; rather, the intelligence rank of the tribes (in descending order: Ami,

Rukai, Puyuma, Tsou, Atayal, Paiwan, Saisiat, Yami, and Bunun) is in general agreement with the civilization ranks given by anthropologists. In other words, degree of intelligence seems more or less a measure of civilization than of genetic differences. There may be genetic variations in general mentality and in special faculties among the tribes, but social components in the variation far outweigh the genetic. It may be assumed that natural selection is mainly directional in favor of high intelligence in such less civilized groups, since to survive depends not only upon physical fitness but also upon intelligence.

Our anthroposcopic data showed variations (Chapter 5) that cannot be associated with any particular environmental factors but are the consequences of evolution, with natural selection and genetic drift playing the most important roles. Consistent sexual variations were also revealed, some of which are not due to sexual selection. For example, in fingerprint patterns our data showed that men have relatively more whorls and loops, and fewer arches, than women, and that these differences are consistent in right hands and throughout the tribes (Chapter 9). In this connection it should be noted that bimanual differences also are consistent throughout the tribes, right hands having more whorls and loops, and less arches, than do left hands. Since ridge formation in hands is completed in the fourth month of pregnancy and does not change thereafter, hormonal differences play no role here. It therefore appears that sexual differences in the frequency of fingerprint patterns reflect differences in the growth and development of the hands between the sexes.

In the above discussion of evolutionary mechanisms involved in tribal differentiation, we emphasized the role of natural selection. The large mortality and fertility rates of the aborigines and the fact that some tribes, such as the

Paiwan and Yami, have had static populations for the past 50 years (Chapter 4) are other indications of natural selection. This emphasis should not be construed, however, as excluding mutation and genetic drift as two other important factors in tribal variations. In natural populations, such as the Taiwan aboriginal tribes, it is impossible to separate these mechanisms, since there are so many unknown variables. But it is generally recognized that, as a basic force in promoting evolutionary changes, mutation supplies raw genetic materials for the free disposal of natural selection. It perpetuates and multiplies those genes that bring about better fitness to the organisms. In recent Drosophila experiments (Mukai, 1964) there has been evidence of a higher mutation rate in polygenes than in major genes. Further understanding of the genetics of evolution awaits our further understanding of the genetics of continuously varying characters.

As between the usefulness of polymorphic and polygenic traits in evaluating relationships between isolated populations, and disregarding the adaptive differences between them, polymorphic traits provide the more precise information, but polygenic traits (e.g., physical measurements), cover many more genes. Statistical technics such as the discriminant function and Mahalanobis' D^2 analyses make it possible mathematically to combine information regarding polygenic traits to express relative relationships between several populations. Thus a great step forward is provided in the use of physical measurements. The high efficiency of combining a large number of polygenic traits for estimating the genetic relationship between parent and offspring, for example, has been shown by Keiter (1963). In comparing the value of polygenic and polymorphic traits in studying population dynamics the problem becomes statistical as in our evaluation of the relationships between the Taiwan

Table 58. Distribution of ABO blood group frequencies in the Taiwan aboriginal tribes (Kutsuna and Matuyama, 1939).[a]

Tribe	Blood groups			
	O	A	B	AB
Atayal	43.0	31.2	20.7	6.0
Saisiat	37.9	30.3	24.6	7.2
Bunun	44.4	36.8	14.7	3.8
Tsou	57.4	28.3	11.6	1.2
Paiwan	66.0	13.7	24.4	1.9
Rukai	52.3	10.1	35.0	2.5
Puyuma	44.8	18.7	31.0	5.4
Ami	35.8	31.0	24.5	8.5
Yami	42.6	39.9	12.6	4.9

[a] The sample sizes for the different tribes range from about 270 in Rukai to 14,000 in Ami.

aboriginal tribes, in which the data for head and body measurements and for dermatoglyphics apparently supplied more information than those for polymorphic traits, such as PTC taste threshold (Chapter 7) and ABO blood group (Table 58, Kutsuna and Matuyama, 1939). It is interesting that the contrast between the Ami and the other tribes was reflected in the result of our discriminant function and distance analyses based on polygenic traits, whereas no clear-cut difference of any kind between the Ami and the other tribes was indicated by the findings based on the frequencies of polymorphic traits.

Polygenic traits, owing to their large stored genetic variability, together with the pleiotropism and linkage of the genes, have flexible properties in response to environmental changes, as discussed in the introduction. Such genetic properties are important in safeguarding the population from environmental changes and in providing a slow and smooth shift. Some of the correlations we found between head and between body measurements (Chapter 6) are not surprising, most of the measurements having been taken either from adjacent or functionally related parts of the same skeleton. But the correlation between measurements of the head and those of the body are interesting, since they relate to entirely different bones. Such correlations are evidently due to the effects of pleiotropism and linkage of the genes concerned. In many cases natural selection favors one trait above others in the same genetically correlated direction, so that evolutionary changes are slow. For example, people in the different aboriginal tribes have characteristic forms of head and body that represent genetic changes adaptive to their individual geographic and social environments; yet almost all retain some of the basic features of Mongoloids, such as the brachycephalic head, prominent cheek bones, some instances of epicanthi, dark hair, iris color, etc. To take another example, American Indians are assumed to have branched out from Mongoloid stock in prehistoric times, but in spite of their long separation they still retain Mongoloid characteristics in physical form and shape.

Because of the large number and the additive properties of genes involved in the polygenic traits we should be cautious in making genetic interpretations. For instance, if two populations have the same mean and distribution of a trait, it does not necessarily follow that they have the same frequencies of the genes concerned. For example, supposing the human head length to be affected by three pairs of genes with equal additive effect but neither dominance nor epistasis, people with the genotype $A_1A_1a_2a_2A_3A_3$, $a_1a_1A_2A_2A_3A_3$, and $A_1a_1A_2a_2A_3A_3$ would be phenotypically similar but genetically different. Nevertheless, when large numbers of traits are included, such as our anthropometric measurements, phenotypic variation is still a good approximation of genetic variation.

Geographic isolation primarily affects traits of adaptive significance. What is an adaptive trait? According to Dobzhansky (1956), it is "an aspect of the whole or of a certain portion of the developmental pattern of the organism. An adaptive trait is, then, an aspect of the developmental pattern which facilitates the survival and/or reproduction of its carrier in a certain succession of environments." On the evidence of our data it seems that some of the variations in head and body forms and measurements among the Taiwan tribes are consequences of adaptation. Allopatric variation brought about by geographic isolation is, therefore, essentially polygenic.

Polymorphic traits are genetically unique because most of them show direct correspondence of characters and genes. But their controlling genetic mechanism is difficult to define, and the factors basic in maintaining their balance are unknown. The best interpretation of the results of our study of the PTC taste threshold in the Taiwan aboriginal tribes is first that the differences in the relative magnitude of dispersion with respect to the different threshold of PTC taste in the different tribes indicate some modifying genes affecting the taste thresholds whose frequency distributions are possibly different among the tribes; and second, that the relatively higher percentages of nontasters occurring in three neighboring tribes, Atayal, Saisiat, and Bunun, indicate either a genetic basis or a biotic environment common to these tribes.

That the genetic mechanisms and factors involved in maintaining the balance of polymorphic traits definitely vary has been discussed in the Introduction. It may generally be said that those of no adaptive significance physiologically can be considered "satellites," varying according to the trait or traits that are adaptive through linkage or pleiotropism, their balance being associated with the composition of the gene pool of the population. Such traits may be neutral physiologically, but not genetically; their variation is still an expression of some characteristic. In effect, the similar percentages of nontasters of PTC in the Atayal, Saisiat, and Bunun, and of ABO blood group types in the Paiwan, Rukai, and Puyuma tribes agree with the results of our anthropometric study. Thus, if more polymorphic traits were studied, it might not be surprising if the total findings for the two sets of traits, polygenic and polymorphic, were in closer agreement as regards the tribal relationships.

Human groups are unusual natural populations in that they have special social structures and partially create their own environment, two factors that complicate genetic studies. Our evidence is not adequate to provide a full explanation of the mechanisms of tribal differentiation or of their genetic relationships. But if the data collected from these marginal populations, and my method of approach, should suggest some fresh concept for the consideration of geneticists and anthropologists, I shall be well satisfied with my study.

BIBLIOGRAPHY

Items marked with an asterisk are not referred to in the text.

Aird, I., H. H. Bentall, and J. A. R. Roberts, "A relationship between cancer of stomach and the ABO blood groups," *Brit. Med. J. 1*, 797–801 (1953).

Anthropology Department of Empirical Taiwan University, *The Taiwan Native Tribes: A Genealogical and Classificatory Study* (1935); in Japanese.

Bailey, D. W., "A comparison of genetic and environmental influences on the shape of the axis in mice," *Genetics 41*, 207–222 (1956).

Bajema, C. J., "Estimation of the direction and intensity of natural selection in relation to human intelligence by means of the intrinsic rate of natural increase," *Eugenics Quart. 10*, 175–187 (1963).

Barclay, G. W., *A Report on Taiwan Population to the Joint Commission on Rural Reconstruction 1954* (Office of Population Research, Princeton University, 1954).

Barnicot, N. A., "Taste deficiency for phenylthiourea in African Negroes and Chinese," *Ann. Human Genet. 15*, 248–254 (1951).

Best, C. H., and N. B. Taylor, *The Physiological Basis of Medical Practice* (Baltimore: Williams & Wilkins, 1950).

Birdsell, J. B., "Some implications of the genetical concept of race in terms of spatial analysis," *Cold Spring Harbor Symp. Quant. Biol. 15*, 259–314 (1950).

* Blackwell, R. Q., and J. T. H. Huang, "Abnormal haemoglobin studies in Taiwan aborigines," *Science 139*, 771–772 (1963).

Bodmer, W. F., and P. A. Parsons, "Linkage and recombination in evolution," *Advances in Genetics 11*, 2–87 (1962).

Bøe, J., S. Humerfelt, and F. Wedervang, "The blood pressure in a population," *Acta. Med. Scand. (Suppl.) 157*, 321 (1956).

Bonnevie, K., "Studies on papillary patterns of human fingers," *J. Genet. 15*, 1–111 (1924).

Boyd, W. W., "Three general types of racial characteristics," *Cold Spring Harbor Symp. Quant. Biol. 15*, 233–242 (1950).

Chai, C. K., "Analysis of quantitative inheritance of body size in mice: I. Hybridization and maternal influence," *Genetics 41*, 157–164 (1956).

221

—— "Analysis of quantitative inheritance of body size in mice: II. Gene action and segregation," *Genetics 41,* 165–178 (1956a).

—— "Analysis of quantitative inheritance of body size in mice: IV. An attempt to isolate polygenes," *Genet. Res. (Cambridge) 2,* 25–32 (1961).

—— and M. S. M. Chiang, "The inheritance of careener, unbalanced locomotion in mice," *Genetics 47,* 435–441 (1962).

* Chan, C. W., "Anthropological studies on the Bunun in Tai-Ping Village, Hau-Lien Prefecture, Taiwan," *Quart. J. Anthropol. 7,* 857–888 (1960); in Japanese.

* Chang, C. S., "Anthropological studies of the skulls of Atayals in Formosa," *Bull. Anat. Dept. National Taiwan University, Taiwan 6,* 59 (1949); in Japanese.

Chang, C. Y., *Atlas of the Republic of China,* vol. I: Taiwan (The National War College in Cooperation with the Chinese Geographical Institute, 1959).

* Chang, E. H., "Anthropological studies on the Atayal in the Hsiu-Lin District, Formosa," *Quart. J. Anthropol. 7,* 919–944 (1960); in Japanese.

* Chang, P. L., and S. Y. Liao, "Palmar dermatoglyphics in the Atayal in Tung-Shyh District, Taiwan," *Bull. Dept. Archaeol. 12,* 35 (1958).

Chang, Y. C., *A Comparative Name-List of Pei-Po Fan's Village* (Historical Bulletin II, No. 1–2, Taipei: The Historical Research Commission of Taiwan, 1951); in Chinese.

Chen, C. L., Y. Y. Li, and M. H. Tang, "Preliminary report of ethnological investigations in the Thao of Jih Yueh Tan (Sun-Moon Lake)," *Taiwan Bull. Dept. Archaeol. Anthropol. 6,* 26–33 (1955).

Chen, Chi-lu, "Family and marriage of the Budai Rukai of Ping-Tung," *Taiwan Bull. Ethnolog. Soc. China 1,* 103–123 (1955); in Chinese with English summary.

Chen, C. S., *A Geography of Taiwan,* vol. I (Taipei: Fu-Ming Geographical Institute of Economic Development, 1960); in Chinese.

—— and C. T. Divan, *The Population of Taiwan* (Bank of Taiwan, 1951); in Chinese.

Chen, H. G., *The Brief History of Immigration in Taiwan: Taiwan Culture (1)* (China Culture Publishing Commission, 1954); in Chinese.

* Chen, I. L., P. Y. Hzieh, and C. W. Chen, "Palmar dermatoglyphics of the Atayal in Chin Shui Tsun, Misoli Hsian, Taiwan," *Bull. Dept. Archaeol. Anthropol., National Taiwan University, Taipei, Taiwan 13–14,* 67–76 (1959).

* Chen, L. T., T. P. Hsu, and C. K. Lin, "Palmar dermatoglyphics in the Vataan Ami, Taiwan," *J. Formosan Med. Assoc. 61,* 853–867 (1962).

* Cheng, F., C. Chang, C. C. Chen, and H. Rin, "A comparative investigation of Paiwanese and urban people by Rorschach test," *Essays and papers in memory of late President Fu Siu-Nien, National Taiwan University* (1962).

Chetverikov (Tschetwerikoff), S. S., "On certain features of evolutionary process from the viewpoint of modern genetics," *J. Exp. Biol.* 2, 3–54 (1926): in Russian; translation in *Proc. Am. Philos. Soc.* 105, 167–195 (1961).

Chouke, K. S., "The epicanthus or Mongolian fold in Caucasian children," *Am. J. Physiol. Anthropol.* 13, 255–279 (1929).

Chow, L. P., and S. C. Hsu, "Taiwan's population problem," *Pop. Rev.* 4, 17–36 (1960).

Clarke, C. A., et al., "The relationship of the ABO blood groups to duodenal ulcer and gastric ulceration," *Brit. Med. J.* 2, 643–646 (1955).

Comas, J., *Manual of Physical Anthropology* (Springfield: Thomas, 1960).

Coon, C. S., *The Origin of Races* (New York: Knopf, 1962).

Cruz-Coke, R., R. Etcheverry, and R. Nagel, "Influence of migration on blood pressure of Easter Islanders," *Lancet 1,* 697–699 (1964).

Cummins, H., and C. Midlo, *Finger Prints, Palms and Soles* (New York: Dover, 1961).

—— and M. Steggerda, "Finger prints in a Dutch family series," *Am. J. Phys. Anthropol.* 20, 19–41 (1935).

Darlington, C. D., and K. Mather, *The Elements of Genetics* (London: Allen & Unwin, 1949).

Department of Civil Affairs, Taiwan Provincial Government, *Household Registration Statistics of Taiwan 1959–1961* (1963).

—— Unpublished statistics (1961).

Dobzhansky, Th., "A review of some fundamental concepts and problems of population genetics," *Cold Spring Harbor Symp. Quant. Biol.* 20, 1–15 (1955).

—— "What is an adaptive trait?" *Am. Naturalist 90,* 337–347 (1956).

—— *Mankind Evolving* (New Haven: Yale University Press, 1962).

Doyle, M. F., and J. R. E. Fraser, "Essential hypertension and inheritance of vascular reactivity," *Lancet 2,* 509–511 (1961).

Falconer, D. S., *Introduction to Quantitative Genetics* (New York: Ronald Press, 1961).

Fisher, R. A., "The use of multiple measurements in taxonomic problems," *Ann. Eugen. 7,* 179–188 (1936–37).

Ford, E. B., *Ecological Genetics* (New York: Wiley, 1964).

—— "Polymorphism and taxonomy," in J. Huxley, ed., *The New Systematics* (Oxford: Clarendon Press, 1940), 493–513.

Fortuyn, A. B. D., "Formation of single-ovum twins," *Quart. Rev. Biol. 7,* 298–306 (1932).

Fuller, J. L., and W. R. Thompson, *Behavior Genetics* (New York: Wiley, 1960).

Galton, F., *Finger Prints* (London: Macmillan, 1895).

Garn, S. M., A. B. Lewis, and J. H. Vicinus, "Third molar polymorphism and its significance in dental genetics," *J. Dent. Res. 42,* 1344–1363 (1963).

Garvin, J. B., "Hypotension," *J. Am. Med. Assoc. 88*, 1875–1876 (1927).

Gates, R. R., *Human Genetics* (New York: Macmillan, 1946).

Grahn'en, H., "Hereditary factors in relation to dental caries and congenitally missing teeth," in C. H. Witkep, Jr., ed., *Genetics and Dental Health* (New York: McGraw-Hill, 1961).

Grüneberg, H., "Die Vererbung der Menschlichen Tastfiguren," *Z. Induktive* Abstammungs Vererbungslehre 46, 285–310 (1928).

—— "Quasi-continuous variation in the mouse," *Symposia Genetica, University of Pavia, Italy*, vol. 3, 215–227 (1952).

Haldane, J. B. S., "The comparative genetics of color in rodents and carnivora," *Biol. Rev. 2*, 199–212 (1925).

Hanna, B. L., "The biological relationship among Indian groups of the Southwest: analysis of morphological traits," *Am. J. Phys. Anthropol. 20*, 499–508 (1962).

Harris, H., and H. Kalmus, "The distribution of taste threshold for phenylthiourea of 284 sib pairs," *Ann. Eugen. (London) 16*, 226–230 (1952).

—— "The measurement of taste sensitivity to phenylthiourea (P. T. C.)," *Ann. Eugen. (London) 16*, 226–230 (1949).

Harrison, G. A., "Genetic bases of morphological variation, by R. H. Osborne and F. V. De George," *Am. J. Phys. Anthropol. 19*, 213–215 (1961).

Hermann, L., and L. Hogben, "The intellectual resemblance of twins," *Proc. Roy. Soc. (Edinburgh) 53*, 105–129 (1933).

Holt, S. B., "Genetics of dermal ridges: maximization of intraclass correlation for ridge-counts," *Ann. Eugen. 17*, 293–301 (1953).

* Hong, Y. T., "Anthropological studies on the Rukai tribe in Tanan District, Tai-Tung, Taiwan," *Quart. J. Anthropol. 7*, 678–708 (1960); in Japanese.

* Hou, Y., "Anthropological studies on the Atayal in the Kappan District," *Quart. J. Anthropol. 2*, 181 (1955).

* —— "Dermatoglyphics of the Atayal in the Kappan District, Taiwan," *Quart. J. Taiwan Museum 9*, 87–107 (1956).

* —— and S. Hu, "Anthropometric measurements of the Atayal in the Gaogan District, Formosa," *J. Formosan Med. Assoc. 52*, 519 (1953).

Howells, W. W., "Factors of human physique," *Am. J. Phys. Anthropol. 9*, 159 (1951).

* Hsi, P. C., "Somatological studies on the Eastern-Paiwan tribe in Ta-Jen District, Tai-Tung Prefecture, Taiwan," *Formosan Science 14*, 249–285 (1960).

* —— "On the hair vortex of the Eastern-Paiwan tribe in Ta-Jen District, Tai-Tung Prefecture, Taiwan," *Formosan Science 14*, 287–295 (1960).

* —— "On the comparison in length between the index and ring fingers of the Eastern-Paiwan tribe in Ta-Jen District, Tai-Tung Prefecture, Taiwan," *Formosan Science 14*, 297–303 (1960).

* Hsieh, P. Y., "An anthropological study of the Bunun in Cho-Chi Village, Hau-Lien Prefecture, Taiwan," *Quart. J. Anthropol.* 7, 889–918 (1960); in Japanese.

* Hsu, T. P., "Anthropological studies on the Vataan Ami, Hua-Lien Prefecture, Formosa," *Nagasaki Igakkai Zasshi 37,* 382–402 (1962); in Japanese.

* Hu, S., "Anthropological studies on the Atayal in the Maliqoan District, Formosa," *Quart. J. Anthropol. 3,* 37 (1956); in Japanese.

Huang, S. L., C. S. Jsai, and P. L. Chou, "On the hair vortex of the Rukai tribe in Mau-Lin District, Kaohsiung Prefecture, Taiwan," *Taiwan Med. J. 5,* 773–778 (1958); in Chinese.

* Hung, Y. T., and T. Tsai, "Blood groups of the Tali-Peipos in the Hua-Lien Province, Formosa," *Quart. J. Anthropol. 6,* 1085–1089 (1959); in Chinese.

Hurst, C. C., "A genetic formula for the inheritance of intelligence in man," *Proc. Roy. Soc. (London) 112,* 80–97 (1932).

—— "The genetics of intellect," *Eugen. Rev. 26,* 33–45 (1934).

Huxley, J., *Evolution* (New York: Harper, 1943).

Information Bureau of the Taiwan Provincial Government, *Reconstruction of Taiwan* (1962).

* Ino, Y., "Are the Botel Tobago aboriginals Negroes?" *Tokyo Zinruigakukai Zasshi 23,* 307 (1908); in Japanese.

* Jou, T. C., "On the hair vortex of the Bunun tribe in San-Ming District, Kaohsiung Prefecture, Taiwan," *J. Formosan Med. Assoc. 59,* 536–542 (1960); in Chinese.

—— and T. M. Wang, "On the blood pressure of the Raval Subtribe of Paiwan Tribe in San-Ti District, Ping-Tung Prefecture, Taiwan," *Quart. J. Anthropol. (Japanese) 7,* 531–545 (1960).

—— and T. M. Wu, "On the blood pressure of the Peipo tribe residing in the Chia-Sien District, Kaohsiung Prefecture, Taiwan," *Quart. J. Anthropol. (Japanese) 7,* 546–560 (1960).

* Kanaseki, T., and Y. Nakano, "On a complete set of a male skeleton of the Botel Tobago aboriginals," *Zinruigaku Zasshi 45,* 1, 182, 200 (1930); in Japanese.

* —— J. K. Tseng, and C. S. Chang, "Kraniometrie des Paiwan und des Bunun-Schädel, Formosa," *Bull. Anatom. Dept. National Taiwan University, Taiwan 6,* 205–216 (1949); in Japanese.

* —— T. L. Tsai, and C. S. Chang, "Anthropological studies of the aboriginals of the island Botel Tobago," *Bull. Anatom. Dept. National Taiwan University, Taiwan 6,* 165–195 (1949); in Japanese.

* —— T. L. Tsai, and C. S. Chang, "Some Yami skulls, vertebras and pelvic bones," *Bull. Anatom. Dept. National Taiwan University, Taiwan 6,* 155–164 (1949).

Kano, T., *Outline Review of the Taiwan Archeology and Ethnology* (The Historical Commission of Taiwan Province, 1955); translated from Japanese into Chinese by W. H. Sung.

Kao, T. Y., *Taiwan Historical Events* (Taipei: Cheng Chung Publishing House, 1958); in Chinese.

Kean, B. H., "The blood pressure of the Cuna Indians," *Am. J. Trop. Med.* *24*, 341–343 (1944).

Keiter, F., "The impact of normal on clinical genetics in man," *Proceedings of the Second International Congress on Human Genetics, 1961* (Rome: Istituto G. Mendel, 1963), pp. 500–503.

Kendall, M. G., *A Course in Multivariate Analysis* (London: Griffin, 1957).

Kimura, K., "The Aimes, viewed from their finger and palm prints," *Z. Morphol. Anthropol. 52*, 178–198 (1962).

* Kirihara, S., and R. Han, "The relations between the tribes of the South Sea Islands and the Formosan aboriginal on the point of blood group studies," *Aichi Igaku Zasshi (J. Aichi Med. Assoc.) 35*, 2169 (1936); in Japanese.

Krauss, W. W., "Eyes of the Pacific," *Asia 42*, 218–220 (1942).

Kroeber, A. L., *Anthropology* (New York: Harcourt, Brace, 1948).

Ku, C. W., F. M. Cheng, F. Y. Huang, and T. M. Lin, *Elementary School Intelligence Test Report on the Non-Verbal Scale* (Republic of China: Dept. Elementary Education, Ministry of Education, 1961).

* Kudo, M., "The Botel Tobago aboriginal viewed from the dactyloglyphical point," *J. Formosan Med. Assoc. 27*, 899 (1928); in Japanese.

* Kutsuna, M., "The physical characteristics of the Formosan aboriginals," *Taiwan Ziho (The Periodical of Formosa) 235,* 116 (1936); in Japanese.

—— and M. Matuyama, "On the blood groups of the Taiwan aboriginal tribes," *J. Taiwan Med. Assoc. 38*, 1153–1178 (1939); in Japanese.

* —— K. Tokitsu, and S. Arizumi, "Plantar epidermal ridge configurations of the Yami tribes on the island Botel Tobago, Formosa," *J. Formosan Med. Assoc. 40*, 506 (1941); in Japanese.

Landauer, W., "Recessive and sporadic rumplessness of fowl: Effects on penetrance and expressivity," *Am. Naturalist 89*, 35–38 (1955).

Lasker, G. W., "Genetic analysis of racial traits of the teeth," *Cold Spring Harbor Symp. Quant. Biol. 15*, 191–203 (1950).

Li, C. C., *Population Genetics* (Chicago: University of Chicago Press, 1955).

* Lin, H. S., C. H. Lai, and C. L. Want, "Palmar dermatoglyphics in the Alisan Tsou, Taiwan," *Quart. J. Taiwan Museum 10*, 71–84 (1957).

Ling, C. S., *The Ancient Min Yuei Tribe and the Taiwan Indigenous Tribes: Taiwan Culture*, 1 (China Culture Publishing Commission, 1954).

Lyon, W., "Zigzag: A genetic defect of the horizontal canals in the mouse," *Genet. Res. 1*, 189–195 (1960).

* Mabuchi, T., "Retrospect on the classification of the Formosan aborigines," *Japan. J. Ethnol. 18*, 1–11 (1953); in Japanese.

* MacKay, G. L., *From Far Formosa* (New York: Revell, 1896).

McKusick, V. A., "Genetics and the nature of essential hypertension," *Circulation 22*, 857–863 (1960).

Maddocks, I., "Possible absence of essential hypertension in two complete Pacific Island populations," *Lancet 2*, 396–399 (1961).

Mahalanobis, P. C., "Analysis of race mixture in Bengal," *J. Asiatic Soc. Bengal 23*, 301–333 (1925).

—— D. N. Majumdar, and C. R. Rao, "Anthropometric survey of the United Provinces, 1941: A statistical study," *Sankhya 9*, 90–324 (1949).

Majumdar, D. N., and C. R. Rao, *Race Elements in Bengal* (London: Asia Pub. House, 1960).

Martin, R., *Lehrbuch der Anthropologie* (Jena: Gustav Fischer, 1928), vol. 1.

* Maruyama, Y., "On the hygienical conditions of the Yami tribes on the island of Botel Tobago. 1. On the general conception of the Yami tribes," *J. Formosan Med. Assoc. 34*, 1962 (1935); in Japanese.

* —— and T. Wakeshima, "The blood groups of the Formosan aboriginals. 4. On the blood groups of the Yami tribes," *J. Formosan Med. Assoc. 34*, 1185 (1935); in Japanese.

Master, A. M., C. I. Garfield, and M. B. Walters, *Normal Blood Pressure and Hypertension* (London: Henry Kimpton, 1952).

Mather, K., "Polygenic inheritance and natural selection," *Biol. Rev. 18*, 32–64 (1943).

* Matsumura, A., "A preliminary report on the aboriginals of Formosa," *Zinruigaku Zasshi 44*, 257 (1929); in Japanese.

Mayr, E., *Animal Species and Evolution* (Cambridge: Harvard University Press, 1963).

Miall, W. E., and P. D. Oldham, "The hereditary factor of arterial blood pressure," *Brit. Med. J. 1*, 75–80 (1963).

* Miyauchi, E., "The stature and the cephalic form of the Formosan aboriginals," *The Thirtieth Anniversary Memoirs of the Government Museum of Formosa* (1938), p. 337; in Japanese.

* Montagu, M. F. A., "A consideration of the concept of races," *Cold Spring Harbor Symp. Quant. Biol. 15*, 315–336 (1950).

—— *An Introduction to Physical Anthropology* (Springfield, Illinois: Thomas, 1951).

* Mori, U., "Formosan aborigines," in *Nippon Encyclopaedia* (Tokyo, 1912); in Japanese.

Mourant, A. E., *The Distribution of Human Blood Groups* (Oxford: Blackwell, 1954).

—— "Human blood groups and natural selection," *Cold Spring Harbor Symp. Quant. Biol. 24*, 57–63 (1959).

Mukai, T., "The genetic structure of natural populations of Drosophila melanogaster. 1. Spontaneous mutation rate of polygenes controlling viability," *Genetics 50*, 1–19 (1964).

Murrill, R. I., "A blood pressure study of the natives of Ponape Island, Eastern Carolines," *Human Biol. 21*, 47–57 (1949).

Newman, H. H., F. N. Freeman, and K. L. Holzinger, *Twins: A Study of Heredity and Environment* (Chicago: University of Chicago Press, 1937).

* Ohashi, H., T. Edamatsu, K. Tokotsu, S. Kaku, and K. Kitagawa, "Odontological studies of the Formosan aboriginals. 1. Yami tribe (Botel Tobago Island). 2. On the teeth caries in the Yami tribe," *Nippon Kokugakukai Zasshi 34*, 151 (1942); in Japanese.

Osborne, R. H., and F. V. De George, *Genetic Basis of Morphological Variation* (Cambridge: Harvard University Press, 1959).

—— F. V. De George, and J. A. L. Mathers, "The variability of blood pressure: basal and casual measurements in adult twins," *Am. Heart J. 66*, 176–183 (1963).

Pearson, E. S., and H. G. Hartley, *Biometrika Tables for Statisticians* (Cambridge: Cambridge University Press, 1956).

Pickering, G. W., *High Blood Pressure* (New York: Grune & Stratton, 1955).

Pickford, R. W., "The genetics of intelligence," *J. Psychol. 28*, 129–145 (1949).

Platt, R., "Heredity in hypertension," *Lancet 1*, 899–903 (1963).

Rao, C. R., "The concept of 'distance' between two groups," in R. Mukherjee, C. R. Rao, and J. C. Trevor, eds., *The Ancient Inhabitants of Jebel* (Cambridge: Cambridge University Press, 1955).

—— *Advanced Statistical Methods in Biometric Research* (New York: Wiley, 1952).

Reed, S. C., "The evolution of human intelligence: some reasons why it should be a continuous process," *Am. Scientist 53*, 317–326 (1965).

Reeve, E. C. R., "Some genetic tests on asymmetry of sternopleural chaeta number in Drosophila," *Genet. Res. (Cambridge) 1*, 151–172 (1960).

* Rin, H., "An investigation into the incidence and clinical symptoms of mental disorders among Formosan aborigines," *Psychiat. Neurol. Japon. 63*, 480–502 (1960); in Japanese.

* —— and T. Y. Lin, "Mental illness among Formosan aborigines as compared with the Chinese in Taiwan," *J. Mental Sci. 108*, 134–146 (1962).

Robinson, S. C., and M. Brucer, "Range of normal blood pressure: statistical and clinical study of 11,383 persons," *Arch. Int. Med. 64*, 409–444 (1939).

Saldanha, P. H., and J. Nacrur, "Taste thresholds for phenylthiourea among Chileans," *Am. J. Phys. Anthropol. 21*, 113–119 (1963).

Scheinfeld, A., *The New You and Heredity* (Philadelphia: Lippincott, 1950).

Schmalhausen, I. I., *Factors of Evolution: The Theory of Stabilizing Selection* (Philadelphia: Blakiston, 1949).

Schwab, R. H., D. L. Curb, J. L. Matthews, and V. E. Schulze, "Blood pressure response to a standard stimulus in the white and Negro races," *Proc. Soc. Exp. Biol. Med. 32*, 583–585 (1935).

Schweitzer, M. D., E. G. Clark, F. R. Gearing, and G. A. Perera, "Genetic factors in primary hypertension and coronary artery disease: reappraisal," *J. Chron. Dis. 15*, 1093–1108 (1962).

Sebesta, P., and V. Lebzelter, "Anthropological measurements in Semangs and Sakais in Malaya (Malacca)," *Anthropologia 6*, 3–4, 183–254 (1928).

Shields, J., *Monozygotic Twins* (London: Oxford University Press, 1962).

* Shima, Y., "Anthropological studies on the Formosan aboriginals. 1. On the dermatoglyphics of the aboriginal on the Pacific coast of Formosa," *J. Formosan Med. Assoc. 38 (Suppl.)*, 1–403 (1939); in Japanese.

Snedecor, G. W., *Statistical Methods* (Ames: Iowa State College Press, 1957).

* Soda, N., "On the hygienical conditions of the Yami tribe on the island of Botel Tobago. 2. On the population and the physical characteristics of the Yami tribes," *J. Formosan Med. Assoc. 34*, 1964 (1935); in Japanese.

* —— "On the physical characteristics of the Yami tribes on the island Botel Tobago," *J. Formosan Med. Assoc. 34*, 2175 (1935); in Japanese.

Stern, C., *Human Genetics* (San Francisco: Freeman, 1949).

* Suzuki, S. T. R., *A Study of the Taiwan Aboriginal Tribes* (Taipei: Historical Research Commission of Taiwan, 1932); in Japanese.

* Taiwan Sotokufu, "Takasagozoku-chosasho (Reports of investigation of the Formosan aborigines," (Taiwan Taihoku, 1938); in Japanese.

Talor, L. W., U. K. Abbott, and C. A. Gunns, "Further studies on diplopodia. 1. Modification of phenotypic segregation ratios by selection," *J. Genet. 56*, 161–178 (1959).

Terman, L. M., and M. H. Oden, *The Gifted Group at Mid-Life* (W. Palo Alto: Stanford University Press, 1959).

* Tokitsu, K., "On the mouth, lips, and occlusion forms of the Yami tribes on the island Botel Tobago, Formosa," *J. Formosan Med. Assoc. 40*, 2256 (1941); in Japanese.

* Torii, R., "To what kind of a race belong the aboriginals of the island Botel Tobago?" *Chigaku Zasshi (J. Geol.) 10*, 419 (1898); in Japanese.

* —— "The photographical picture series from the Anthropological Department of the Tokyo Imperial University: The Botel Tobago Island" (1899), with a somatological explanation in Japanese.

* —— "The cephalic form of the Botel Tobago aboriginals," *Tokyo Zinruigakukai Zasshi (J. Anthropol. Soc. Tokyo) 16*, 308 (1901); in Japanese.

* —— "The stature and the span of arms of the Botel Tobago aboriginals," *Tokyo Zinruigakukai Zasshi (J. Anthropol. Soc. Tokyo) 17*, 85 (1901); in Japanese.

* —— "Études anthropologiques: Les aborigènes de Formose (2ᵉ fascicule),"
  J. Coll. Sci. Tokyo Imperial University 32, 1–75 (1910).
* —— "Études anthropologiques: Les aborigènes de Formose (2ᵉ fascicule).
  1. Caracters physiques: Une tribu Yami," (Tokyo: Tokyo Imperial
  University, 1922).
Tryon, R. C., "Genetic differences in maze-learning ability in rats," in
  39th Yearbook Nat. Soc. Study Educ. (Bloomington, Illinois: Public School
  Pub. Co., 1940), Part I, 111–119.
* Tsai, C. S., "Somatological studies on the Atayal tribe in the Gaogan
  District, Tao-Yuan Prefecture, Taiwan," *Studia Taiwanica 3*, 1–30
  (1959).
* Tsai, T. L., T. C. Jou, and C. S. Tsai, "Somatological studies on the Rukai
  tribe in Mau-Lin District, Kaohsiung, Taiwan. The First Report:
  Somatoscopy," *Taiwan Med. J. 57*, 702–709 (1958); in Chinese.
* —— T. C. Jou, C. S. Tsai, and S. C. Lee, "Somatological studies on the
  native tribe residing in Mantauran Village of Mau-Lin District (The
  Rukai Tribe), Kaohsiung, Taiwan," *J. Formosan Med. Assoc. 58*, 80–92
  (1958).
* —— T. C. Jou, C. S. Tsai, and S. C. Lee, "Somatological studies on the
  Rukai tribe in Mau-Lin District, Kaohsiung, Taiwan. The Second
  Report: Somatometry," *J. Formosan Med. Assoc. 58*, 211–223 (1959); in
  Chinese.
* Tseng, T. M., "Anthropological studies on the Bunun (Take-Vataans) in
  Ma-Yen Village, Hua-Lien County, Taiwan (Formosa)," *Memoirs of
  the College of Medicine, National Taiwan University 6*, 17–56 (1960).
U. S. Bureau of the Census, Department of Commerce, *U. S. Census of
  Population* vol. 1 (1960).
Waddington, C. H., *The Strategy of the Genes* (New York: Macmillan,
  1957).
—— H. Graber, and B. Woolf, "Iso-alleles and response to selection,"
  J. Genet. 55, 246–250 (1957).
* Wang, F. P., "On the comparison in length between the index and ring
  fingers of the Raval Subtribe of Paiwan Tribe in San-Ti District,
  Ping-Tung Prefecture, Taiwan," *Quart. J. Anthropol. 7*, 795–800 (1960);
  in Japanese.
Wang, S. H., "Family structure of the Kahiyagan Paiwan," *Bull. Dept.
  Archaeol. Anthropol. 13–14* (1959).
Washburn, S. L., "The strategy of physical anthropology," in A. L.
  Kroeber, ed., *Anthropology Today* (Chicago: University of Chicago Press,
  1953), 714–727.
* Wei, H. L., "Lineage system among the Formosan tribes," *Bull. Inst.
  Ethnol., Academia Sinica 5*, Taipei, Taiwan (1958).
—— *The Origin and Classification of the Taiwan Indigenous Ethnic Groups in
  Taiwan Culture (1)* (China Culture Publishing Commission, 1954); in
  Chinese.

Whitney, D. D., *Family Treasures: A Study of Inheritance of Normal Characteristics in Man* (Lancaster: Cattell Press, 1942).

Wilder, H. H., "Scientific palmistry," *Pop. Sci. Monthly 61,* 41–54 (1902).

Williamson, M. H., "On the polymorphism of the moth *Panaxia dominula,*" *Heredity 15,* 139–151 (1960).

* Woo, Y. C., "On the comparison in length between the index and ring finger of the Mu-Cha Peipo in Nie-Men District, Kaohsiung Prefecture, Taiwan," *Quart. J. Anthropol. 7,* 637–642 (1960); in Japanese.

* —— "On the hair vortex of the Mu-Cha Peipo in Nie-Men District, Kaohsiung Prefecture, Taiwan," *Quart. J. Anthropol. 7,* 709–716 (1960); in Japanese.

Wright, S., "Evolution in Mendelian populations," *Genetics 16,* 97–159 (1931).

—— "General, group, and special size factors," *Genetics 17,* 603–619 (1932).

—— "An analysis of variability in number digits in an inbred strain of guinea pigs," *Genetics 19,* 506–536 (1934).

—— "The genetics of quantitative variability," in E. C. R. Reeve and C. H. Waddington, eds., *Quantitative Inheritance* (London: H. M. Stationery Office, 1952).

* Wu, T. C., C. W. Chen, and E. H. Chang, "Palmar dermatoglyphics in the Atayal in Li-Shan Village, Hua-Lien Prefecture, Formosa," *Quart. J. Anthropol. 7,* 711–727 (1960); in Japanese.

* Wu, T. M., "Somatological studies on the Hsin-Fua Peipo in Liu-Kuei District, Kaohsiung Prefecture, Taiwan," *Quart. J. Anthropol. 7,* 608–636 (1960); in Japanese.

* —— "On the hair vortex of the Peipo Tribe residing in the Hsin-Fua Village of Liu-Kuei District, Kaohsiung Prefecture, Taiwan," *Quart. J. Anthropol. 7,* 757–764 (1960).

* Yang, C., and T. Li, "Über das hautleistensystem der planta des Bununstammes in Formosa," *Bull. Anatom. Dept. National Taiwan University 3,* 332 (1948); in Japanese.

* Yen, H., "Dermatoglyphics of the Atayal in the Kappan District, Taiwan," *Quart. J. Taiwan Museum 9,* 87–107 (1956).

* Yen, I., and H. Lin, "The personality evaluation of the Yami tribe inhabiting Botel Tobago Island by Rorschach test," *Bull. Ethnolog. Soc. China 1,* 217 (1955).

* Yoneyama, T., and T. Wakeshima, "Psychomedical studies on the Formosan aboriginals. Statistical observation on the psychoanomalies in the Formosan aboriginals," *J. Formosan Med. Assoc. 40 (Suppl.),* 1–31 (1941); in Japanese.

* Yu, C. C., and T. M. Cheng, "Physical anthropology of the Thao, Sun-Moon Lake," *Bull. Dept. Archaeol. Anthropol. National Taiwan University 9–10,* 125–136 (1957); in Chinese.

* Yu, Y. H., *Journeys in the Formosa Straits* (1697; republished by Historical Research Commission of Taiwan Province, Taipei); in Chinese.

INDEX

Aboriginal tribes: visited in field survey, 16; history, classification, and distribution, 24–37; lineage, 29–33; excluded from field survey, 32–33; extinct, 33–34; civilization ranks, 34; geographic and topographic distribution, 34–36; population statistics, 38–41; age distribution, 41, 43; birth, death, and fertility rates, 43; marriage customs, 43–44; inbreeding, 44; family structure, 44, 45–47; succession in, 48; Mongoloid characteristics, 77, 218; correlations between anthropometric measurements of, 97, 102–103; relationship to other ethnic groups, 112–113; biologic relationships, 118, 185; chi-square test for fingerprint variations among, 162–163; eastern and western, 163; basis of variation among, 209, 211–212; correlation of tribal variations and relationships with geographical and social isolation, 210; ancestry, 210; intelligence ranks, 214–215

Adaptive traits, 96, 145, 213, 219

Allopatric variation, 219

Ami tribe: association with Puyuma, 32; civilization rank, 34; geographic and topographic distribution, 34; population size and growth, 39, 40–41; marriage customs, 44; household size, 45; succession in, 48; anthroposcopic observations, 51, 53, 56, 57, 59, 77, 78; anthropometric measurements, 84, 88, 93; biologic separation, 88, 116, 117, 118; homogeneity, 112; PTC taste threshold, 119, 122, 125,

126; blood-pressure test, 133, 134, 139, 140, 214; correlation of blood pressure and chest girth, 144; dermatoglyphics of fingers and palms, 159, 161, 164, 173; intelligence test, 191, 196, 200, 205; correlation of geographical with biological isolation, 211; physical characteristics, 211; intelligence rank, 214–215

Anthropology, physical, 1, 2

Anthropometric measurements: and correlation between polygenic characters, 13; number taken on aborigines, 79; procedures, 79–81, 83–84; symbols used in recording, 84; head indices, 90–94; body indices, 94–95; tribal variation, 95–96; stature of Bununs and Negritos, 96; chest-girth variations, 96–97; correlations between measurements, 97–103; intertribal variance, 103–108

Anthropometry, decline in use of, 1

Anthroposcopic observations: of mucous lips, 49–50, 51, 77; of nasal-bridge forms, 51, 52, 53, 77; of eye apertures, 53, 54, 77; of eyefolds, 54–55, 56, 57; of iris color, 59, 77; of ear lobes, 60–61, 77; of ear points, 62, 77; of hair color and form, 64, 77; of skin color, 64, 77; difficulty of classifications, 75; sexual variations, 215; variations, evolutionary, 215

Atayal tribe: association with Saisiat, 32; civilization rank, 34; geographic and topographic distribution, 34; population size and growth, 39–40; marriage customs, 43; succession in,

233

Atayal tribe (cont.):
47; anthroposcopic observations, 53, 57, 58, 61, 62, 77–78; biologic distances, 88, 115, 116, 117; anthropometric measurements, 93, 94–95; classification, 112; PTC taste threshold, 121, 122, 124, 125, 126, 219; genetic association, 128; dermatoglyphics of fingers and palms, 161, 172, 180–182; heterogeneity, 212; intelligence rank, 215

"Baldwin effect," 213
Biologic separation of tribes, 212
Blood groups, 9, 10, 220
Blood pressure: hypertension, 130, 131, 132, 145; hypotension, 131; of U.S. industrial employees, 132; average, for American male adult, 132; of people in Bergen, Norway, 132–133; regression on age, 132–133, 134, 139; mean age of aborigines tested, 133; auscultatory method, 133; tribal means, 139–141; sex differences, 141; Hunter's observations, 141; averages compared, 141–142; effect of body build, 144; effect of inbreeding, 144; effect of natural selection, 145; predisposition of Mongoloids to hypotension, 144, 146; genic theories on variation, 145–146; genetically less definable, 214; influence of Mongoloid genetic background, 214
Body size as polygenic character, 11, 12
Bunun tribe; intelligence test, 19, 200, 201, 205; geographic and topographic distribution, 34; civilization rank, 34; population size and growth, 39, 40; marriage customs, 43; household size, 45; pedigree data, 46, 48, succession in, 48; anthroposcopic observations, 53, 59, 64, 76, 77–78; infiltration by Negritos, 76–77, 96, 210; anthropometric measurements, 84–88, 93, 94–95; biologic separation, 88, 116, 117, 118, 212; chest girth, 96, 213; homogeneity, 112; classifi-

cation, 112; PTC taste threshold, 122, 124, 125, 126, 219; genetic association, 128; blood pressure, 134, 144; correlation of blood pressure with chest girth, 144; dermatoglyphics of fingers and palms, 159, 181; intelligence rank, 196; physical characteristics, 211; geographic isolation correlated with biologic, 211–212

Caucasoids, 210
Chi-square test: percentage of probability considered significant, 51; of relation of ear point to ear lobe, 63–64; of fingerprint-pattern differences, 151–152, 162–163; of bimanual and sexual differences in palmar main lines, 168–169, 170; of heterogeneity, 202
Civilization: effect on relation of age to blood pressure, 132; correlated with intelligence, 215
Correlation: between fingerprints and palm lines, 13; between anthropometric measurements, 97, 102–103; between head and body measurements, 102; of biologic relationships with geographic distribution, 118; between nontasters and geographic distribution, 126; between body build and blood pressure, 144; between ridge count and pattern intensity, 163; between digits in fingerprint patterns, 175–178; between palmar main lines, 177; between ridge pattern and function, 184; between geographical and biological isolation, 211; between biologic and anthropologic findings, 212; between dermatoglyphic and anthropometric variations, 213; between intelligence and civilization, 215

Data from field survey: scope, 14; technics used to analyze, 14
Dermatoglyphics: correlation between fingerprints and palm lines, 13;

Dermatoglyphics (cont.):
history, 147–148; heritability, 147–148; as quantitative genetic trait, 148; three-factor hypothesis, 148; fingerprint patterns, 148, 149, 150, 151–152, 159–164, 175, 176, 182, 183, 215; ridge-pattern formation, 148, 149, 177–178; technics for analyzing variations, 149; finger and palm prints, 149–150; Galton's four categories, 149; fingerprint-pattern intensity, 149, 164, 165; hand and finger differences, 150–151; sexual differences in fingerprints, 151–152; bimanual and sexual differences, 151–152, 164, 168, 183, 215; asymmetry, 159; tribal differences in fingerprints, 159–162; ratio of whorls to loops, 163–164; correlation of pattern intensity and ridge count, 165; palmar main lines, 165–169, 172–174, 183, 184; thenar and hypothenar patterns, 169; correlation between digits in fingerprint patterns, 176; correlation of palmar main lines, 177; genetics, 177–178; and natural selection, 179–180, 215; Mahalanobis' D^2 analysis, 180–182; fingerprints and tribal relationships, 182; variation in A line, 183; palmar main lines in primates, 184; correlation of ridge pattern with function, 184; correlation between variations and anthropometric measurements, 213
Discriminant function analysis: of anthropometric measurements, 108–113; of relative tribal heterogeneity, 114; of dermatoglyphic data, 178; efficiency of technic demonstrated, 211
Drift, genetic, 216

Environment: as limitation of population increase, 41; effect, 107, 108; as component in between-tribe variance, 108; and PTC taste threshold, 127; and blood pressure, 132; and

sexual differences in intelligence, 205; and natural selection, 213
Evolution: mechanisms, 5, 6, 216; genetic bases, 8–14; and polygenic variations, 11; maintenance of original forms, 13, 14; of ear point and ear lobe, 63; and sexual variations in fingerprints, 158; affected by functional differences, 184; and variation in blood pressure, 214; and continually varying characters, 216; and natural selection, 218

Field survey, 15–23
Formosa. *See* Taiwan

"Gene pool": defined, 3; relation to polymorphic traits, 9, 219; and polymorphism, 10; and equilibrium of PTC genes, 128
Genes: and pleiotropy, 3; physiologic properties and evolution, 3; polygenes, 4; flow between tribes, 112–113; additive properties, 218; theory of supergenes, 8, 9
"Genetic assimilation," 213
Genetic component in anatomic characters, 213
"Genetic load," 5–6
Genetics, Mendelian: and study of natural populations, 2; PTC taste threshold, 126; of dermatoglyphics, 177–178; of development, 213
Geographic isolation and adaptive traits, 219

Hair forms, 11
Heredity, in intelligence, 186–187
Heritability: of stature and correlated body measurements, 106; of hypertension, 130–131; of dermatoglyphics, 147–148
Hyperplasia. *See* Blood pressure: hypertension

Infiltration: of Bununs and Paiwans by Negritos, 77; between Puyuma and Rukai tribes, 212

Inheritance, polygenic, 11

Intelligence: as polygenic variation, 11; in monozygotic and dizygotic twins, 186; role of heredity, 186–187; strain or breed differences, 186–187; genetics, 187; purpose of testing aborigines, 187–188; correlated with civilization, 203–204; sexual differences, 205; influence of environment, 205–206; effect of natural selection, 206; correlation with family size, 206–208; correlated with special faculties, 208; and natural selection, 208; difficulty of defining genetically, 214; tribal ranks, 214–215; and civilization, 215; social and genetic components, 215

Intelligence tests: described, 188, 190; procedure, 188–190; mean ages of children tested, 190–191; test for normality, 191–194; distribution patterns, 193–196; correlation between parallel forms, 194; age regression on scores, 195–196; sexual differences, 196; variation in tribal mean scores, 200; comparative reliability, 205

Isolation: geographic and social, basic to tribal variations, 118, 209; "river and shore line territory," 211

Linkage: as controlling mechanism of polymorphism, 10; concerning polygenic characters, 13; effect on anthropometric correlations, 218

Mahalanobis' distance analysis: of anthropometric measurements, 113–118; compared with discriminant function analysis, 114; value in determining biologic separation of tribes, 114, 209, 216; of dermatoglyphic data, 180–182

Mendelian population, genotypes in, 8

Mongoloids, 210

Mutation, 5, 216

Natural selection: "Darwinism reborn," 2; types, 5, 213; as evolutionary mechanism, 5, 215; and "genetic load," 5–6; in favor of heterozygotes, 8, 9; and polymorphic variation, 8, 9; and polygenetic theory, 11; response of pleiotropism and linkage, 13; effect on chest girth of Bununs, 96; physical and dermatoglyphic characters, 179–180, 214; effect on traits induced by environment, 213; favors one trait above others, 218

Negritos: disappearance, 33–34; possible infiltration of Bununs, 76–77, 210; mean stature compared with Bununs', 96; head form and stature discussed by Kroeber, 102; earliest inhabitants of Taiwan, 209, 210

Nontasters, tribal averages, 124–125

Paiwan tribe: association with Rukai, 32, 118, 212; geographic and topographic distribution, 34; civilization rank, 34; population size and growth, 39, 41, 216; marriage customs, 43; succession, 48; anthroposcopic observations, 51, 53, 54, 61, 64, 78; anthropometric measurements, 84–88, 93, 94–95, 96; biologic separation, 88, 116, 118; classification, 112; PTC taste threshold, 122, 125; blood pressure, 134; dermatoglyphics, 161, 175–176, 177, 181; intelligence test, 191; physical characteristics, 211; intelligence rank, 215

Panmictic unit, 3, 7

Ping Pu tribe, 32–33, 141

Pleiotropism: significance in evolution, 4; as controlling mechanism of polymorphism, 10; as property of polygenes, 13; as property of taster and nontaster genes, 128; effect on anthropometric correlations, 218

Polygenic traits: genetic complexity, 2; variations, and evolution, 10–14; adaptability, 11, 217; examples, 11, 12, 13; homeostasis and flexibility, 13; correlations between, 13; compared to polymorphic traits, 216–217, 220

Polygenes: defined, 4; as determinants of quantitative characters, 4; additive effects, 13; properties, 13, 218

Polymorphic traits: genetic uniqueness, 1, 219; information supplied by, 216; compared to polygenic, 216; "satellites," 219

Polymorphism: defined, 8; supergenes in maintenance of, 8, 9; and natural selection, 8, 9; variations, 8–10; adaptive significance, 9, 10; controlling mechanisms, 10

Population genetics, 1, 2; correlations of polygenic traits, 13

PTC taste sensitivity: method of testing, 119–120; nontasters defined, 124n; single-gene hypothesis, 126; shift in antimode, 126; tribal percentages of nontasters correlated geographically, 126; and goiter, 126–127; interpretation of test data, 219

Puyuma tribe: geographic and topographic distribution, 34; civilization rank, 34; population size and growth, 39, 41; marriage customs, 43; succession in, 48; anthroposcopic observations, 51, 53, 57, 59, 77; biologic separation, 88, 116, 118; anthropometric measurements, 93, 94–95; association with Rukai, 118; PTC taste threshold, 119, 123, 125; blood pressure, 133, 134, 140, 141, 142, 144; dermatoglyphics of fingers and palms, 159, 163, 173, 181; intelligence test, 191, 200; intelligence rank, 215

Rukai tribe: geographic and topographic distribution, 34; civilization rank, 34; population size and growth, 39, 41; marriage customs, 43–45;

intermarriage, 44; family types, 45; succession in, 48; anthroposcopic observations, 51, 53, 56, 61, 64; anthropometric measurements, 84–88, 93, 94–95; classification, 112; biologic separation, 116, 118; association with Paiwan, 118, 212; PTC taste threshold, 122, 125; blood pressure, 134, 140; dermatoglyphics, 159, 164, 169, 173, 182; contact with Puyumas as western group, 163–164, 169; intelligence test on, 196–199; physical characteristics, 211; intelligence rank, 215

Saisiat tribe: civilization rank, 34; geographic and topographic distribution, 35; population size and growth, 39, 40–41; marriage customs, 43; succession in, 48; anthroposcopic observations, 51, 53, 61, 62, 77; anthropometric measurements, 93, 94–95; classification, 112; biologic separation, 116, 117; PTC taste threshold, 120, 122, 124, 125, 126, 219; genetic association, 128; blood pressure, 133, 134, 139, 140, 141, 142, 144; dermatoglyphics, 159, 160, 169, 172, 173, 181; intelligence test, 196, 200; intelligence rank, 215

Sexual selection, 6

Skin color, 11

Social selection, 6

South Pacific Islands, 210

Taiwan, 24–26; first mention of indigenous people, 27–28; racial framework, 210

Tawu Mountain, 161

Tsou tribe: civilization rank, 34; geographic and topographic distribution, 35; population and rate of growth, 39, 41; household size, 45; succession in, 48; anthroposcopic observations, 51, 53, 54, 57, 78; anthropometric measurements, 88, 93, 94–95; classi-

Tsou tribe (cont.):
fication, 112; PTC taste threshold, 122, 125; blood pressure, 133, 134; dermatoglyphics, 160, 163, 179, 182; intelligence test on, 196; intelligence rank, 215

Twin studies: by Newman et al., 106; by Osborne and DeGeorge, 106–107; compared to studies of human populations, 106–108

Yami tribe: inaccessibility, 20, 37; population size and growth, 39, 41, 216; succession in, 48; intelligence test, 191, 196; intelligence rank, 200, 215